THE TIMES
UNIVERSE

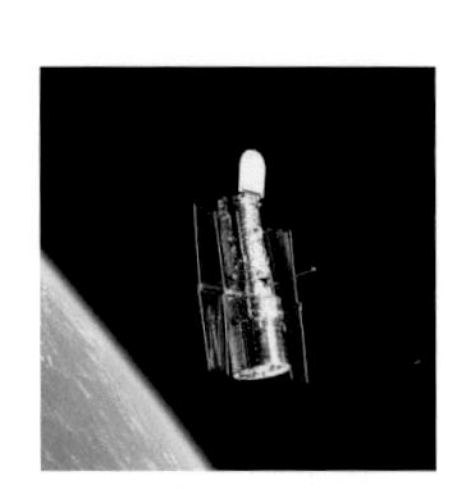

THE TIMES UNIVERSE

IAN RIDPATH

First published in 2002 as *Times Space* by TIMES BOOKS
This edition published 2004

HarperCollins*Publishers*
77-85 Fulham Palace Road
London
W6 8JB

The HarperCollins website address is:

www.**fire**and**water**.com

ISBN 0 00 716930 2

Design and layout by Ian Ridpath

Colour origination by Colourscan, Singapore

Printed and bound by Printing Express Ltd., Hong Kong

Contents

Introduction

Sunset over the dome of the 4.2-metre William Herschel Telescope on La Palma in the Canary Islands – the prelude to another night's observing.

When the Italian scientist Galileo Galilei turned his first primitive telescope on the night sky nearly 400 years ago he saw unsuspected wonders. Among them were mountains and craters on the surface of our Moon, phases of the planet Venus, four moons orbiting Jupiter, and countless faint stars beyond the compass of the naked eye. He immediately recognized that the existing picture of the Universe, necessarily based on human vision alone, was very incomplete.

Galileo's findings revolutionized the way that humans perceived their celestial environment, yet they were made with a telescope that, by modern standards, was only a crude toy. As telescopes improved and enlarged, astronomers were able to peer ever deeper into cavernous space, mapping a cosmic hierarchy of stars, galaxies and clusters of galaxies in which our home planet Earth came to seem increasingly insignificant. Today's instruments can show the Universe as it appeared at an era close to Time Zero, long before the Earth was born, when space and time are believed to have come into being in a primeval fireball called the Big Bang.

Magical as it may sound to be able to sift through the strata of time to uncover the birth of the Universe, it is an inevitable consequence of the laws of physics. Light and its related messengers – from radio waves at the long-wavelength end of the spectrum to X-rays and gamma rays at the shortest – do not flash instantaneously across space but run to a precise schedule, some 300,000 kilometres per second. The distance covered by a light beam in one year defines the interstellar unit of length, the light year. Hence the farther an object is from us, the more time its light takes to reach us and the more out of date is our view of it. Observing the remote Universe through a large telescope is as near to time travel as we are likely to get.

A telescope's two most vital attributes are its ability to see faint objects and to resolve fine detail. Size is important to achieve these aims, but so also is location. For the clearest, sharpest views of the heavens, astronomers site their telescopes atop high mountains. Prime locations for observatories include Hawaii, La Palma in the Canary Islands and the Andes mountains of South America.

Over the past half century optical telescopes (those that collect visible light) have been joined in the astronomers' armoury by other instruments capable of observing the remainder of the electromagnetic spectrum. First came radio telescopes, developed in the 1950s, followed more recently by satellite observatories in space that can detect infrared, ultraviolet, X-ray and gamma-ray wavelengths that do not penetrate the Earth's atmosphere and so are unobservable from the ground.

A new era began in 1990 with the launch of the Hubble Space Telescope (HST). Its light-collecting mirror is 2.4 metres (94 inches) wide; although smaller than that of the largest telescopes on Earth, it can nevertheless see the sky with far greater clarity and sharpness because it is above the atmosphere. HST's performance was initially impaired because the surface of the mirror was wrongly curved during manufacture, but this was corrected by astronauts who installed additional optics on a servicing visit.

During more than a decade of operation the gaze of HST has ranged from the planets of our Solar System to the edge of the observable Universe. It has shown us close-ups of the collision of a comet with Jupiter, the birth and death of stars, views of the environments of black holes, and snapshots of galaxies in collision. Most significantly of all, HST has allowed astronomers to complete the work begun by the American astronomer Edwin Hubble, after whom the telescope is named, of accurately measuring the size and age of the Universe.

Even HST is far from the ultimate. Already on the drawing board is the James Webb Space Telescope (named after a former NASA administrator), due for launch at the end of this decade with a mirror at least 6 metres (236 inches) diameter, over twice that of the existing Hubble Space Telescope. Its role will be to study the formation of the first stars and galaxies in the early Universe.

HST and its successors-to-come have not rendered ground-based observatories obsolete, though. Astronomers are building a generation of optical telescopes that dwarf the venerable 5-metre (originally known as the 200-inch) reflector on Palomar Mountain, California. Adopting a technique developed by radio astronomers, they are ganging together two or more telescopes to achieve improved performance, as in the twin Keck reflectors in Hawaii, the two-eyed Large Binocular Telescope in Arizona and the four-part Very Large Telescope in Chile.

All these instruments are providing views of the heavens that continue to surprise and astonish. Some of the most sumptuous images taken from Earth and space are collected in this book – yet even these are only a progress report, for the discovery of the Universe continues apace.

◀ **Gang of Four:** The European Southern Observatory's Very Large Telescope (VLT) consists of four individual reflectors each with a mirror 8.2 metres (323 inches) in diameter, atop Paranal mountain in northern Chile. They can be used singly or in unison, in the latter case giving the light-gathering capacity of a single mirror 16 m (630 inches) in diameter. Smaller outlying telescopes will work in conjunction with the larger ones, using a technique known as interferometry to discern smaller details than is possible with the large telescopes alone.

▼ **Living in a box:** The enclosure of the Large Binocular Telescope (LBT) on Mount Graham in Arizona. Uniquely, the LBT consists of two 8.4-m (331-inch) mirrors on a common mounting, and the enclosure has a viewing slit for each mirror covered by sliding doors. Additional openings on the back and sides allow wind ventilation of the building. The LBT will be equivalent in light-gathering power to a single 11.8-m (465-inch) mirror.

New designs are revolutionizing the size and performance of telescopes on Earth. Mirrors 8 metres (315 inches) in diameter, once unthinkable, are now commonplace. To keep the weight of such gargantuan disks of polished glass down to manageable proportions they are made very thin in relation to their diameter. There is a serious disadvantage to this, though: they flex and lose shape. Consequently, the mirrors must be continually adjusted by a complex support system to retain their precise focus as the telescope is moved and tilted. The very largest existing telescopes, the twin 10-m (394-inch) Keck reflectors on the summit of Mauna Kea in Hawaii, have mirrors composed of numerous hexagonal segments which are adjusted individually to maintain a crisp image.

In perhaps the greatest advance of all, cutting-edge technology is being used to overcome the unsteadiness of the Earth's atmosphere which otherwise blurs telescopic images. Before entering the telescope's detectors, the light collected by the main mirror is bounced off a small, pliable mirror which can be warped tens or even hundreds of times a second to counteract the distortions produced by the atmosphere. This technique, called adaptive optics, is capable of producing images of stars which rival those of the Hubble Space Telescope for sharpness.

▲ **Full sweep:** An internal panorama of the 8.1-m (319-inch) Gemini North telescope on Mauna Kea, Hawaii, with the dome closed.

◀ **Open house:** The dome of Gemini North, with the side air vents fully open.

Telescopes have changed outwardly as well as inwardly. Many large modern telescopes are housed not in traditional domes but in cylinders or boxes, giving them the appearance of huge battleship guns. During the day the enclosures are refrigerated to keep the telescope at night-time temperatures, and ventilated at night to prevent air currents that would disturb the steadiness of the image. The most remarkable example is provided by the two 8.1-m (319-inch) Gemini telescopes, one in Hawaii and one in Chile, whose domes have shutters at the sides that open to allow a free flow of air over the telescope during observing (see left). In all modern observatories, the astronomers themselves never enter the dome, but command the telescope from a separate control room.

▶ **Sharpening up:** The central region of the globular cluster NGC 6934 seen through the Gemini North telescope without adaptive optics (left) while, on the right, the smaller area within the white rectangle is shown with adaptive optics switched on. The sharpness of the image on the right is comparable to resolving the separation between two car headlights at a distance of 3000 km (2000 miles).

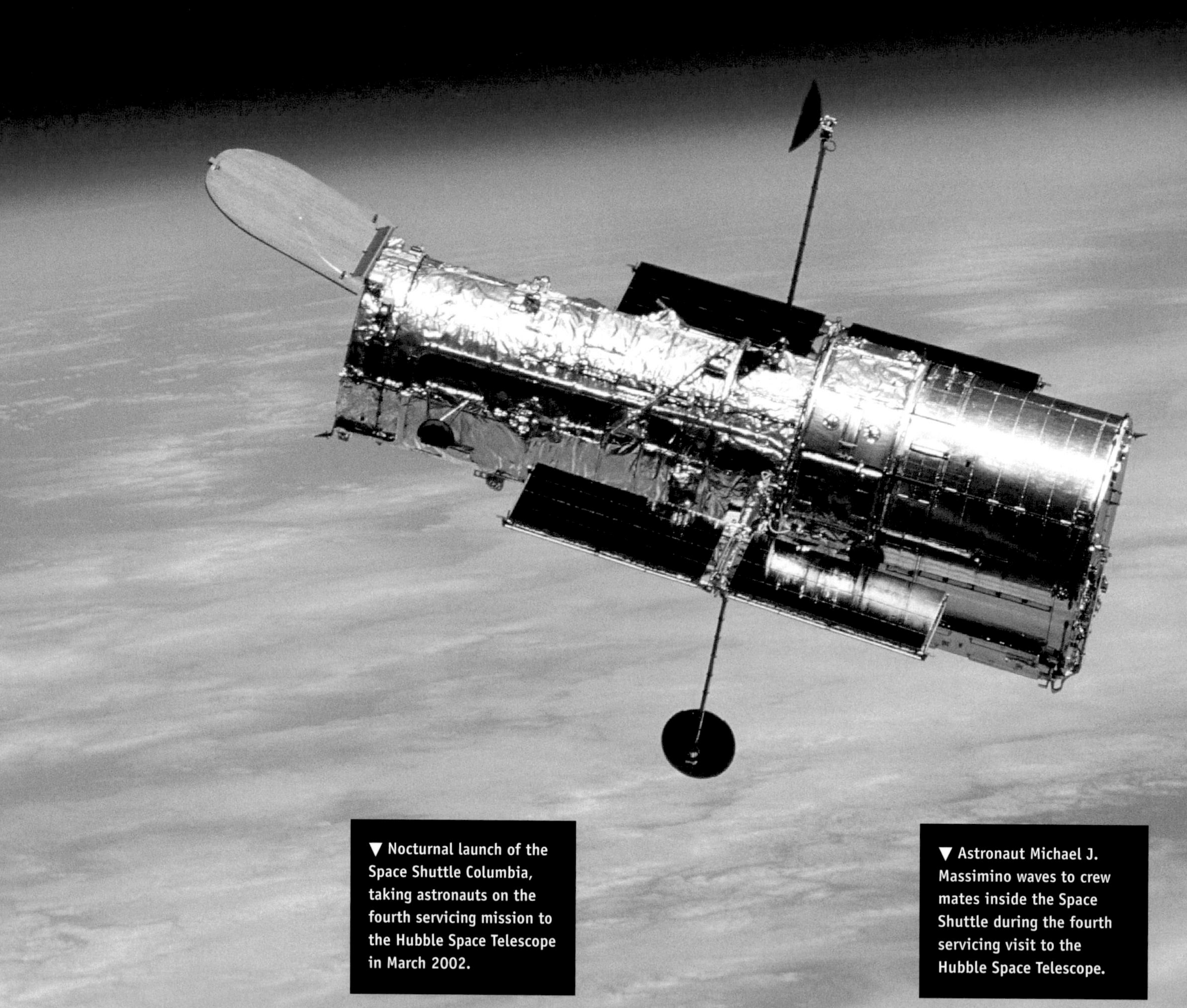

▼ Nocturnal launch of the Space Shuttle Columbia, taking astronauts on the fourth servicing mission to the Hubble Space Telescope in March 2002.

▼ Astronaut Michael J. Massimino waves to crew mates inside the Space Shuttle during the fourth servicing visit to the Hubble Space Telescope.

◀ **Silver time machine:** The cylindrical body of the Hubble Space Telescope is flanked by a new set of energy-generating solar panels installed on the fourth servicing mission in March 2002. HST observes the Universe from orbit 600 kilometres (375 miles) above the Earth.

▼ **High achievers:** Astronauts John M. Grunsfeld (top, supported on the end of the Space Shuttle's remote-controlled arm) and Richard M. Linnehan work at the base of the Hubble Space Telescope in the cargo bay of the Space Shuttle Columbia during the fourth maintenance mission in March 2002. The cloud-covered Earth provides a backdrop.

For the first two and half years of its life, the Hubble Space Telescope was unable to focus properly because its main mirror had inadvertently been made with the wrong curvature. Computer processing improved the blurred images, but it was not until the end of 1993, when astronauts installed additional optics and a new camera, that HST's view finally snapped into crisp focus. Additional servicing missions have continued to repair and upgrade HST. On the fourth such mission, in March 2002, astronauts replaced the solar arrays, installed the Advanced Camera for Surveys (ACS) and fitted a new cooling system for Hubble's infrared camera.

Although astronomers are now closing the performance gap on the HST with ever more powerful telescopes on Earth, space-based astronomy will take another leap forward when the James Webb Space Telescope (JWST) is launched around 2010, equipped with a mirror at least 6 m (236 inches) in diameter. The objects that will come within its gaze are so far off that their light has been greatly redshifted by the expansion of the Universe, so they are best observed in the infrared. That requires keeping the telescopes and instruments ultra-cool, to prevent their own infrared emissions from disturbing the observations. JWST will shelter behind a huge sunshade in an orbit 1.5 million km (0.9 million miles) from Earth – some four times the distance of the Moon. Hence service missions like those to HST, shown below, will no longer be possible.

Earth, Moon and Planets

Worlds in space: A montage of the planets, seen over a stark lunar landscape

Our home world, the Earth, is one of a family of nine planets that orbit the Sun, along with a good deal of flying debris. This family, called the Solar System, came into being some 4560 million years ago from a cloud of gas and dust that shrank under the inward pull of its own gravity. Most of the cloud went into the central core, the Sun; the planets were pieced together from the surrounding sweepings.

A cursory glance at the Solar System shows that it divides distinctly into two: an inner half of small, solid planets, with large, gaseous planets in the colder reaches farther from the Sun. Of the four rocky inner planets the largest is the Earth, but only by a small margin over Venus. Our planet is also blessed with an unusually large Moon, fully one-quarter its own size.

The dominant member of the planetary family is Jupiter, a gas-enshrouded ball of liquid hydrogen and helium eleven times the diameter of the Earth. All the other planets combined would still weigh less than half Jupiter. Beyond Jupiter lies the magnificent ringed planet Saturn, then the near-twins Uranus and Neptune. All these giant planets are accompanied by extensive families of moons. Pluto is an icy oddity orbiting at the perimeter of the Solar System, more like an oversized comet than a planet.

In the seam between the inner and outer planets is the ragged band of asteroids, chunks of rocky and metallic rubble ranging in size from Ceres, over a quarter the diameter of our Moon, down to mere pebbles. Inconsequential as they may seem, it would be wrong to dismiss asteroids as of little interest: they carry clues to our origins, for they incorporate samples of the primordial material from which the planets formed. They will also provide a rich source of minerals for future generations.

Some asteroids career drunkenly across the paths of the planets, threatening accidents. One such collision, 65 million years ago, is believed to have done for the dinosaurs, and astronomers now survey the skies for other incoming artillery that could spell our own doom.

Most elegant to behold of all the Solar System's members are the comets, clumps of dust and ice, usually no more than a kilometre or so in diameter, that swarm in a pack called the Oort Cloud at the outer fringes of the Solar System. Occasional gravitational nudges from passing stars send some of these comets plunging in towards the Sun. Warmed by sunlight, they release billowing clouds of gas and dust, becoming temporarily visible to us on Earth. After looping around the Sun they recede into obscurity, not to be seen again in a human lifetime.

Over the past four decades, our view of the Solar System has been transformed. Space probes have scouted all the planets except Pluto and machines from Earth currently rest on the surfaces of two of them, Venus and Mars, as well as asteroid Eros. There are human footprints in the dust of the Moon and lunar rocks on Earth. From indistinct points of light, as they appeared to previous generations, the planets have become individual worlds whose surfaces are more familiar to us than some parts of our own planet.

The illustrations that follow are part of the legacy of our exploration of the Earth and its neighbouring worlds.

Earth, the Blue Marble

Seen from space, our home planet, the Earth, floats in the void like a blue marble, veined with white streaks of cloud. Hurricane Linda rages off the west coast of North America while weather systems in the North Atlantic approach northwestern Europe. Other notable details include the blue-green shallow waters of the Caribbean and yellowish sediments around the mouth of the Amazon River.

The image above was assembled at NASA's Goddard Space Flight Center using data from three different satellites. The clouds were photographed from geostationary orbit by a weather satellite called

GOES (Geostationary Operational Environmental Satellite) on September 9, 1997. The ocean information was assembled from cloud-free scans by the SeaWiFS (Sea-viewing Wide Field-of-view Sensor) satellite. Information on vegetation was collected by the AVHRR (Advanced Very High Resolution Radiometer) instrument on a third satellite. Heavy vegetation is depicted as green while areas of sparse vegetation are coloured yellow. The heights of mountains and depths of valleys have been exaggerated by 50 times so that vertical relief is visible. The presence of the Moon at upper left is an artistic addition; it is not shown to scale.

Stripping away the atmosphere and oceans reveals the solid Earth (above). Shallow continental shelves fringe land masses, most noticeably around North America and northern Europe. Along the west coast of Central and South America are deep ocean trenches where a part of the Earth's crust dips steeply under another and is destroyed. Elsewhere new crust is being extruded, such as along the East Pacific Rise and the mid-Atlantic ridge. This global view was compiled from a combination of information from satellites and ships, and is artificially coloured to indicate altitudes – orange and red are the highest, dark blue the deepest.

▶ **Lighting up:** Artificial lights mark out cities, roads and railways on the Earth at night, as seen from orbit by satellites of the US Defense Meteorological Satellite Program (DMSP). This global mosaic was assembled from scans made by DMSP satellites on cloud-free nights. The eastern United States, western Europe and Japan are swamped with lights, while much of Africa, central Asia, Australia and South America remain dark. Note the Trans-Siberian railroad, a tendril of light stretching from Moscow eastwards across Asia, and the prominent thread of the Nile river in north Africa running south from the Mediterranean Sea to the Aswan Dam. In the United States, a latticework of highways connects the brighter dots of cities. Faced with images such as this, environmentalists have joined astronomers in calling for improved outdoor lighting that conserves energy by minimizing the wasteful spill of light into the sky.

THE DEATH OF NIGHT

Nowhere is the impact of humanity on the Earth better seen than from space, and never is it more noticeable than at night. Scientists use satellite photographs of artificial lights to chart the spread of urbanization, but such photographs also tell another story: that of energy wastage, bathing urban skies in a permanent twilight that obscures our view of the wider Universe we inhabit. Light pollution, as it is termed, is a growing concern to environmentalists as well as astronomers. The photograph at the right, taken from the Space Shuttle, offers a closer view of the bright lights on the northeastern seaboard of the United States, looking obliquely from the north. At left is Long Island and the glare of greater New York. The central splash of light is Philadelphia, with Baltimore and Washington, DC, to the right. Above that is Richmond and, at the top of the frame, Norfolk, Virginia. Many smaller towns are also visible.

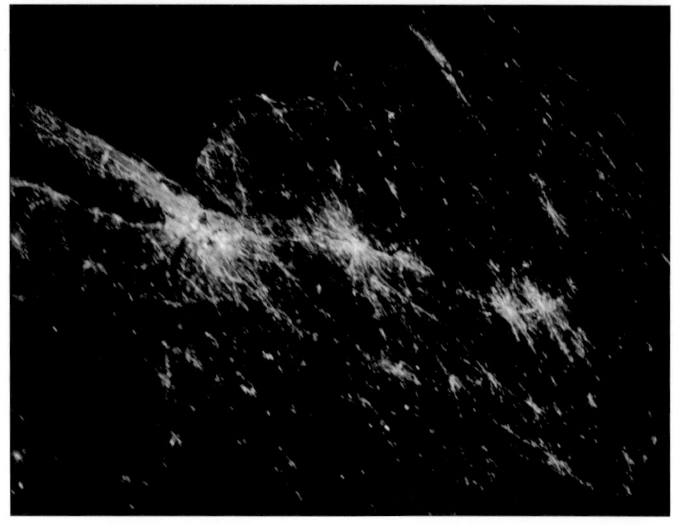

THE OZONE HOLE

Another effect of human activity visible from space is the so-called hole in the ozone layer over Antarctica, which satellites have been monitoring since the 1980s. Ozone in the Earth's upper atmosphere blocks harmful ultraviolet rays from the Sun, but it is destroyed by gases called chlorofluorocarbons (CFCs). The ozone hole, shown in blue at right from readings taken by NASA's Total Ozone Mapping Spectrometer (TOMS), reached its greatest size yet in September 2000, more than 28 million square km (11 million square miles), three times the land area of the United States. Concentrations of CFCs in the stratosphere have started to level off and even decline as a result of restrictions on their use. However, once CFCs reach the stratosphere they can take many years to settle out again. The ozone hole should soon start to shrink, but scientists do not expect ozone levels to return to pre-1980 values for decades yet.

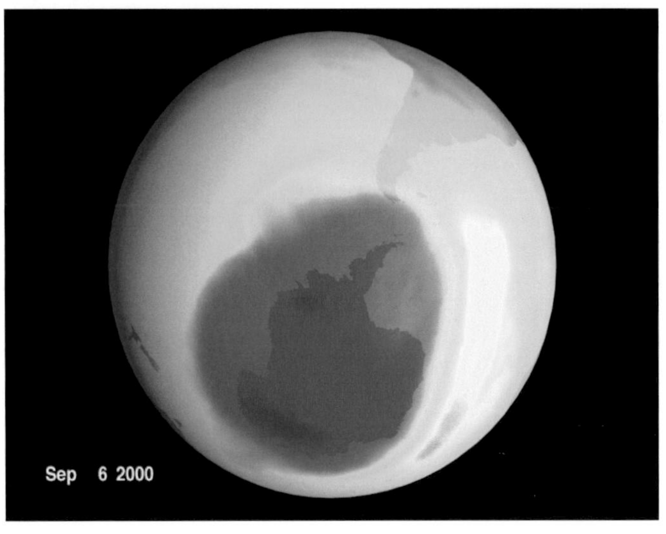

Stepping stone into space

A new star is growing brighter in our skies: the International Space Station (ISS), the largest and most complex outpost ever constructed in space. A joint project of the United States, Russia, Europe, Japan and Canada, the ISS is taking shape nearly 400 km (250 miles) overhead, where it orbits the Earth every hour and a half. Sections of the station are being ferried up by US Shuttles and Russian rockets over a period of several years, although schedules have been delayed by the loss of the Shuttle Columbia in February 2003. When completed, the ISS is planned to have a mass of over 450 tonnes. It will be the size of a football field, with a habitable volume equivalent to three houses or a 747 jumbo jet, and will draw power from multiple sets of wing-like solar panels. The existing US and Russian sections will be joined by a European Space Agency laboratory called Columbus and a Japanese module named Kibo.

The first crew of three, known as Expedition One, took up residence in the partially completed station in November 2000 for four and a half months, and regular crew exchanges have kept the station permanently occupied ever since. Eventually, the ISS will be able to house up to seven people at a time for stays of three to six months.

The ISS opens a new phase in human exploration of space. It is a platform for observation of Earth and sky, a laboratory for experiments in weightlessness, a testbed for new industrial processes and perhaps, in future, a jumping-off point for the Moon and Mars.

◀ **Home delivery:** The Space Shuttle Atlantis approaches the International Space Station in February 2001. The silver cylinder in its cargo bay is the US laboratory module called Destiny, which was added to the ISS. Destiny is used for experiments in the near-zero gravity of space.

▶ **Under construction:** The International Space Station, seen from the Space Shuttle Atlantis after it undocked from the station at the end of assembly mission STS-112 in October 2002. The wing-like extensions at top are the first of an eventual four sets of solar arrays. Protruding beneath the station is a Russian Soyuz craft in which the crew can return to Earth in an emergency.

MIR – THE FORERUNNER

Before the International Space Station there was Mir, a Russian craft launched in February 1986 and subsequently enlarged by various plug-in sections. During its 15-year lifetime Mir was home to more than 60 people from over a dozen countries, one of whom, Valeri Polyakov, set the record for the longest-ever continuous spaceflight, 14 months – sufficient for a trip to Mars. Between 1995 and 1998 nine Space Shuttle missions docked with Mir for joint activities, building up valuable experience for the ISS. Mir's career ended in March 2001 when ground controllers directed it into the atmosphere to burn up over the Pacific Ocean.

▶ **Farewell glance:** Mir seen through the window of the last Space Shuttle to visit it, Discovery, in June 1998.

▼ **Big hit:** A ring of blue water outlines the Manicouagan Reservoir in Quebec, Canada, the eroded remains of a 212-million-year-old meteorite impact. The circular feature, one of the largest impact scars on Earth, is approximately 100 km (60 miles) wide. It was photographed in June 2002 by astronauts on the STS-111 mission which delivered the fifth crew to the ISS.

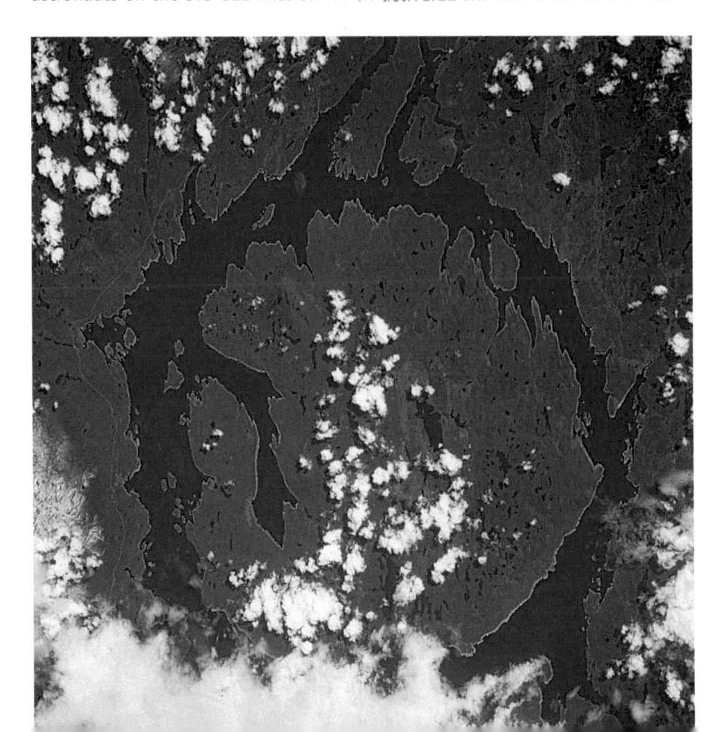

▼ **Smoking:** A plume of dark ash rises from a wintertime eruption of Mount Etna, an active volcano in Sicily, photographed during the STS-113 mission which carried the sixth crew to the International Space Station in November 2002. Etna's snowy peak, over 3300 m (10,900 ft) high, rises through frothing clouds over the Mediterranean Sea in this southward-looking sunset view.

The Moon, our companion

Barren, cratered and coloured only in shades of grey, the Moon would be an unappealing destination for space travellers were it not for the fact that it is by far our nearest neighbour, an offshore island in the interplanetary sea a mere 384,400 km (239,000 miles) from Earth.

The Moon is our constant companion and its influence is pervasive. Its metronomic sweep of phases, from a slim crescent to dazzling fullness and back again in just over four weeks, measures out a natural rhythm in our skies, while its gravity induces the ebb and flow of the tides. In return, the Earth's gravity has put a brake on the rotation of the Moon, locking it into synchronicity with its orbital motion so that one side faces permanently Earthwards.

Even the unaided eye can discern the division of its rocky surface into bright highlands and dark plains, which human imagination assembles into the facial features of the man in the Moon. Binoculars disclose breathtaking detail, particularly along the sunrise–sunset line where low illumination throws mountains and craters into sharp relief.

Ancient impacts, around 4 billion years ago, excavated large basins that later filled with lava to create the dark lunar lowlands. These are termed *maria*, the Latin for "seas", because early astronomers fancifully imagined them to be expanses of water. The charming names remain, such as Mare Imbrium ("Sea of Rains") and Mare Tranquillitatis ("Sea of Tranquillity"), where Apollo 11 landed in July 1969. Craters are named after scientists and philosophers. Little has happened to change the Moon's overall appearance in the past 3 billion years, apart from the occasional arrival of a meteorite to punch a new crater. The youngest craters, such as Copernicus and Tycho (see right), are surrounded by bright rays of debris splashed out by the impacts that created them. Tycho, some 100 million years old, is probably the youngest major feature on the Moon.

Earth and Moon compared

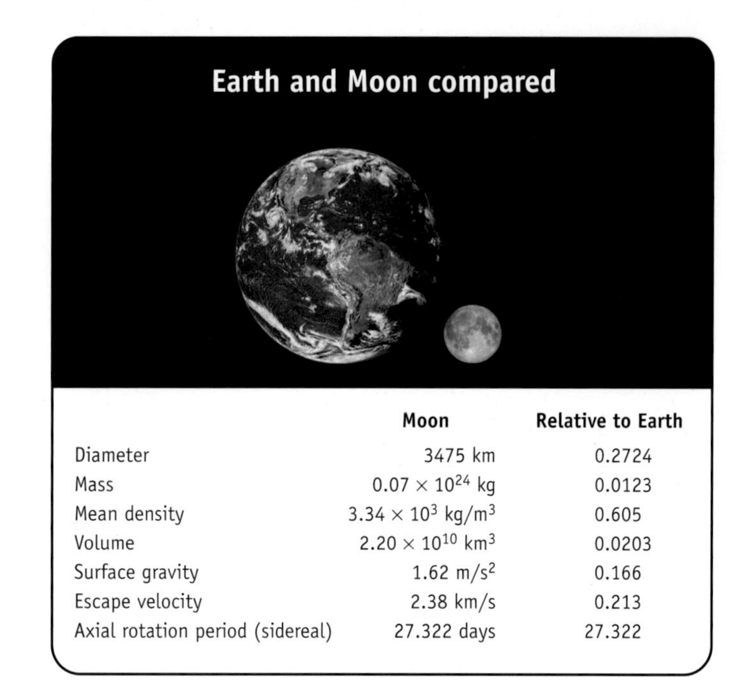

	Moon	Relative to Earth
Diameter	3475 km	0.2724
Mass	0.07×10^{24} kg	0.0123
Mean density	3.34×10^{3} kg/m^3	0.605
Volume	2.20×10^{10} km^3	0.0203
Surface gravity	1.62 m/s^2	0.166
Escape velocity	2.38 km/s	0.213
Axial rotation period (sidereal)	27.322 days	27.322

NEAR AND FAR

The Moon's far side, permanently turned away from our view, remained unknown until space probes circled behind to photograph it. As these mosaics from the lunar-mapping craft Clementine demonstrate, the far side (below) almost entirely lacks the familiar dark lowland plains that mottle the Earth-facing hemisphere (left). The reason is that the Moon's crust is somewhat thinner on the near side, so lava flooded out from the lunar interior to fill the lowlands there.

▶ **Zooming in:** Binoculars bring into view the main features of the Moon, seen here as it appears about four days before full.
Key to the marked formations:
1: Sinus Iridum, a bay flooded by lava from adjacent Mare Imbrium.
2: Plato, a prominent dark-floored crater 100 km wide.
3: Copernicus, a crater 90 km wide encompassed by splashy rays.
4: Mare Crisium, a distinctive lunar "sea" about 500 km wide.
5: Ptolemaeus, a large but shallow depression 150 km wide that becomes more difficult to see as the Sun rises higher over it.
6: Bullialdus, a deep and prominent crater 60 km wide.
7: Tycho, a bright and distinctive crater, the centre of an extensive system of rays.
8: Clavius, a beautiful crater 225 km wide with a chain of smaller indentations on its floor.
A 11: The Apollo 11 landing site.

1
2
3
4
A 11
5
6
7
8

MAN ON THE MOON

For three and a half glorious years between July 1969 and December 1972, a select band of humans, the Apollo astronauts, realized the dream of generations by walking on the surface of the Moon. While one astronaut remained in lunar orbit in the conical Command Module, two others descended in the lumpy, four-legged Lunar Module. They set foot on a landscape unlike any on Earth: airless, waterless and lifeless, churned into dust by billions of years of exposure to bombardment by meteorites of all sizes.

Using the cramped Lunar Module as a base camp, astronauts stayed for 21 hours on the initial landing, Apollo 11, to over three days on the final one, Apollo 17, which visited the foothills of the Taurus Mountains on the southeastern shores of Mare Serenitatis (Sea of Serenity). From Apollo 15 onwards their explorations were extended by an electrically powered Moon car, the Lunar Rover, stored in the lower half of the Lunar Module. Apollo 17's Eugene Cernan and Harrison Schmitt drove for a record total of 35 km (21¾ miles).

In all, the six successful landings (Apollo 13 was a failure) brought back over 380 kg (840 lb) of rocks and soil. Taking into account the cost of obtaining them, these samples are more valuable than gold.

▲ **The Earth also rises:** The crew of Apollo 8 which orbited the Moon in December 1968 were the first humans to see the far side of the Moon with their own eyes, and to watch the Earth rising over another world in space.

▼ **Stepping out:** Alan Bean carefully makes his way down the ladder of the Lunar Module of Apollo 12, the second lunar-landing mission, which visited Oceanus Procellarum (the Ocean of Storms) in November 1969.

▼ **Made it:** Apollo 11 astronaut Edwin "Buzz" Aldrin stands on the Moon in this classic photograph from the Apollo era, taken by Neil Armstrong. Armstrong and the Lunar Module Eagle are reflected in Aldrin's helmet visor.

▲ **Flying the flag:** James Irwin salutes the American flag outside the Apollo 15 Lunar Module. The mountain in the background is Hadley Delta. Apollo 15 was the first mission to carry a Lunar Rover, seen at right.

Painstaking analysis in laboratories on Earth has teased the secrets of the Moon from these precious specimens. Most astounding is their immense antiquity: of the six sites sampled, the youngest rocks, found by Apollo 12, were laid down around 3.1 billion years ago, comparable with some of the oldest rocks on Earth. The oldest lava flows that fill the maria date back some 3.9 billion years, and the heavily cratered highlands are older still. The Moon is truly a dead world, and has been so for most of its existence.

From this new-found knowledge of the Moon's composition and history, a revolutionary explanation of its origin has emerged. No existing scenario, such as a split from Earth, capture of a passing body or formation alongside the Earth, accorded fully with the facts. Astronomers now theorize that, before the Earth was fully formed, it was struck a glancing blow by a passing body the size of Mars. The incoming body was destroyed in the collision but debris from it, along with some from the outer layers of the embryo Earth, was thrown into orbit where it coalesced into the Moon. Although quiet today, the Moon is a neighbour with a violent past.

▼ **First footing:** Apollo 11 astronaut Buzz Aldrin photographed his own bootprint in the lunar soil. Without rain or wind to disturb them, the footprints of the Apollo astronauts will endure for millions of years.

▲ **Souvenir hunting:** Apollo 12 landed within walking distance of Surveyor 3, a robot probe that had arrived two and a half years earlier as part of a series of exploratory missions to assess the nature of the lunar surface. Here, mission commander Pete Conrad prepares to remove the Surveyor's camera assembly which was brought back to Earth. The Lunar Module is in the background.

◀ **Sideways look:** An oblique view of the crater Copernicus taken from lunar orbit by the Apollo 12 astronauts. From this angle, it is apparent how shallow the crater is in relation to its diameter.

▲ **On the edge:** Apollo 15 astronaut David Scott with the Lunar Rover at the rim of Hadley Rille, a winding valley thought to have been carved out by flowing lava, possibly in a tunnel that later collapsed. This black-and-white photograph accurately conveys the true greyness of the Moon – the tints of the lunar surface seen in most Apollo colour photographs are spurious.

▶ **Ripe orange:** The unremitting greyness of the Moon is relieved in places by some colour, as the Apollo 17 astronauts discovered when they stumbled across this orange soil beside a small crater called Shorty. It consists of glassy droplets probably sprayed out by volcanic lava fountains when the floor of the surrounding Littrow Valley was laid down, 3.7 billion years ago.

◀ **Leaning:** The Apollo 15 Lunar Module, named Falcon, leans at an angle on the Moon's surface, having landed with one leg in a shallow crater. Astronauts' footprints and the tyre tracks of the Lunar Rover criss-cross the soft lunar soil, known as regolith. In the background, the distant Apennine mountains appear nearer than they really are. Only the top half of each Lunar Module returned from the Moon, leaving the lower stage and the Lunar Rover on the surface where they may one day become tourist attractions.

◀ **The scoop:** Harrison Schmitt, the only trained geologist to visit the Moon, scoops up small rocks with a rake from the dark floor of the Littrow Valley during the Apollo 17 mission, the final lunar landing, in December 1972.

▲ **Rock on:** Apollo 17 astronaut Harrison Schmitt is dwarfed by a huge boulder, informally dubbed Tracy's Rock, which has rolled downhill and split. In the distance are the hills of South Massif, about 8 km (5 miles) away across the Littrow Valley.

LUNAR ECLIPSES

While moving along its monthly orbit, the Moon sometimes strays into the shadow cast by the Earth and is eclipsed. This can happen only at full Moon, when Sun and Moon lie opposite each other in our skies, but the alignment must be precise – the Moon's orbit is angled with respect to the Earth's, so usually it ducks our shadow.

As an eclipse progresses, the Earth's shadow spreads across the face of the Moon like a seeping stain until, after more than an hour, the Moon is completely engulfed (except when only the edge of the shadow catches the Moon, in which case the eclipse never becomes more than partial). Unlike at a solar eclipse there is no sudden snapping off of light when totality arrives. Light leaks around the rim of the Earth through its atmosphere and into the shadow, imparting a coppery hue to the Moon which usually remains dimly visible throughout totality. The colour and darkness of the shaded Moon varies between eclipses because of changing conditions in the Earth's atmosphere. After about an hour of totality the Moon begins to re-emerge from the gloom.

At least one total lunar eclipse is visible from any given point on Earth every two or three years, although some of them occur at inconveniently late hours. Many people will never see a total eclipse of the Sun, but everyone should be able to see several total lunar eclipses in their lifetime.

▲ **Chasing shadows:** The Moon arches from horizon to horizon above an eroded rock formation known as the chimney of the fairies during the total lunar eclipse of April 1996. The progress of the eclipse is captured in this montage of images taken at 12-minute intervals from the Bardenas desert in Navarra, northeastern Spain.

▼ **Red, red Moon:** At the lunar eclipse of July 2000 the Moon turned blood red during totality. This multiple exposure shows the extent of the Earth's shadow as the Moon passed through its centre. Totality at this eclipse lasted 1 hour 47 minutes, the longest duration possible.

Mercury, small and rocky

Mercury, the innermost planet, never strays far from the Sun and is difficult to see from Earth. Our only good views of it have come from one space probe, Mariner 10, which flew past it three times in 1974–75 revealing a lunar-like landscape strewn with impact craters from a lifetime of interplanetary bombardment. Such an appearance is no surprise given that Mercury is only 40 per cent larger than the Moon, too small to retain a protective atmosphere. Mercury is in fact smaller than any other planet except Pluto, and smaller even than the largest moons of Jupiter and Saturn.

Mercurial by nature as well as name, the planet speeds around the Sun every 88 days. Yet, in contrast, its axial spin is strangely lethargic – one turn takes nearly 59 Earth days, exactly two-thirds of its orbital period. The curious consequence is that, seen from the surface of Mercury, the Sun requires 176 days to go once around the sky, during which time the planet has orbited the Sun twice and turned on its axis three times with respect to the distant stars.

▼ Cratered world: Mercury's lunar-like surface, in a mosaic of photographs taken by Mariner 10 as it approached the planet in 1974. The bright crater above centre is Kuiper, seen in close-up on the facing page.

Earth and Mercury compared

	Mercury	Relative to Earth
Diameter	4879 km	0.383
Mass	0.33×10^{24} kg	0.055
Mean density	5.43×10^{3} kg/m^3	0.984
Volume	6.08×10^{10} km^3	0.056
Surface gravity	3.70 m/s^2	0.378
Escape velocity	4.30 km/s	0.384
Axial rotation period (sidereal)	58.65 days	58.785
Axial inclination	0.01°	0.0004
Distance from Sun	$46.00 - 69.82 \times 10^{6}$ km	0.307 - 0.467
Eccentricity of orbit	0.206	12.311
Orbital period (sidereal)	87.969 days	0.241

On the Sun-seared dayside, temperatures soar to over 400°C, hot enough to melt tin and lead (were any to be found there), while during the long night the temperature plummets below −180°C. Some permanently shadowed areas near the poles remain forever below the freezing point of water, so ice deposited by impacting comets may survive there.

For such a small planet, Mercury has a surprisingly high density, second only to that of the Earth. This implies that Mercury has a heart of iron – a core that extends out to fully three-quarters its own radius, far greater in proportion than any other planet. Such disproportion smacks of disruption in the past – and indeed a hefty smack from a stray body may well have blasted much of Mercury's original rocky shell into space, leaving it with its current reduced size, as well as an orbit that departs markedly from circularity.

Mariner 10 photographed less than half the planet, so there is much still to learn. Our knowledge of Mercury should improve considerably in the next few years. In 2009 an American probe called Messenger (Mercury Surface, Space Environment, Geochemistry and Ranging) is due to go into orbit around the diminutive world to study its entire surface in detail. It will be followed a few years later by a European orbiter and lander, BepiColombo.

▲ **Splashing out:** Kuiper, a bright impact crater with rays, was one of the most conspicuous features in the first photographs of Mercury returned by Mariner 10 (see opposite). Named after Gerard Kuiper, an American planetary scientist who died while the mission was underway, the crater is 41 km (25 miles) in diameter.

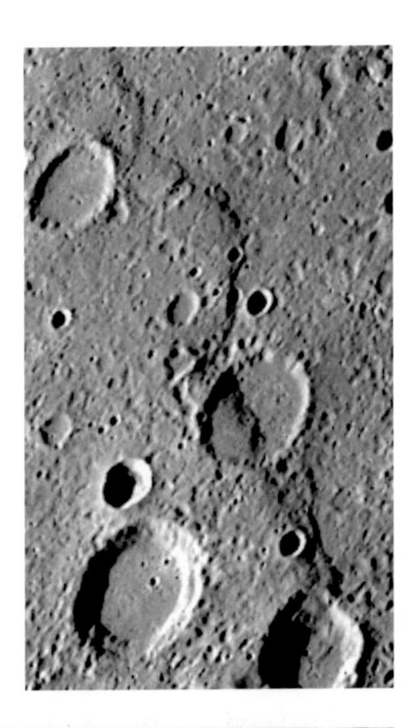

▶ **Shrink wrapped:** One type of feature unique to Mercury is meandering faults in the crust known as lobate scarps. These are thought to result from a slight shrinkage of the planet as it cooled after formation. This scarp, Discovery Rupes, is about 500 km (300 miles) long and 1 km high, and cuts through several craters.

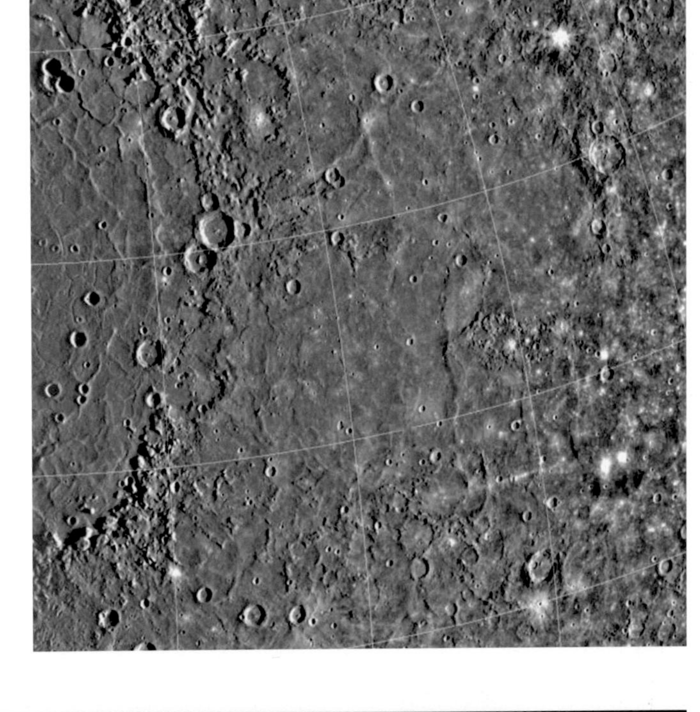

▶ **Big basin:** The largest feature on Mercury is the Caloris Basin, 1300 km (800 miles) wide, an impact scar larger even than Mare Imbrium on the Moon. Only the eastern side of the huge formation, at the left of this image, was photographed by Mariner 10; the remainder was in darkness at the time.

CROSSING THE SUN

On rare occasions, Mercury can be seen through telescopes as a tiny black dot silhouetted against the enormity of the Sun's globe, an event known as a transit. For a transit to occur, Mercury must firstly be at a point where it crosses the plane of the Earth's orbit, which happens in May and November. In addition, it must lie directly between the Earth and Sun. The most recent transits were in 1999, when Mercury just clipped the Sun's northern edge (see photograph), and 2003. The next is in 2006 on November 8, visible from North and South America, Australasia, Japan and east Asia. After that there will not be another transit of Mercury for nearly a decade.

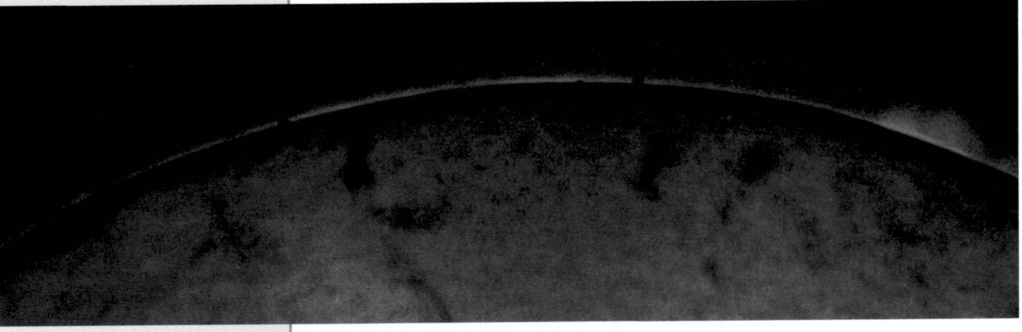

▶ **In transit:** Multiple exposures of Mercury appear like a row of perforations as the planet clips the Sun's rim at the transit of 1999 on November 15. The images of Mercury in transit were combined with an exposure of the Sun's glowing surface taken in the red light emitted by hydrogen gas.

Venus, cloudy and hot

Venus is unmissable in the sky, shining brilliantly as either the morning or evening 'star', depending on which side of the Sun it lies in its orbit – everyone has seen Venus, even without realizing what it really was. Yet for all its blazing glory to the eye, Venus through a telescope is bland and disappointing. We gaze upon an unbroken cloudscape relieved only by the vaguest of markings, changing in illuminated phase as it orbits the Sun.

In terms of diameter and distance from the Sun, Venus is the planet that most closely matches the Earth, but the two are far from twins. Whereas the Earth has an atmosphere of about four-fifths nitrogen and one-fifth oxygen, shot through with white clouds of water vapour, that of Venus consists almost entirely of unbreathable carbon dioxide, topped with clouds composed of droplets of sulphuric acid, a substance most familiar on Earth as battery acid.

Furthermore, the dense carbon dioxide atmosphere traps heat from the Sun in an extreme demonstration of the greenhouse effect, driving up temperatures to a furnace-like 460°C, from equator to pole, day and night. Needless to say, there is no water on such a hot planet. Compounding this Hellish vision, the atmosphere of

Earth and Venus compared

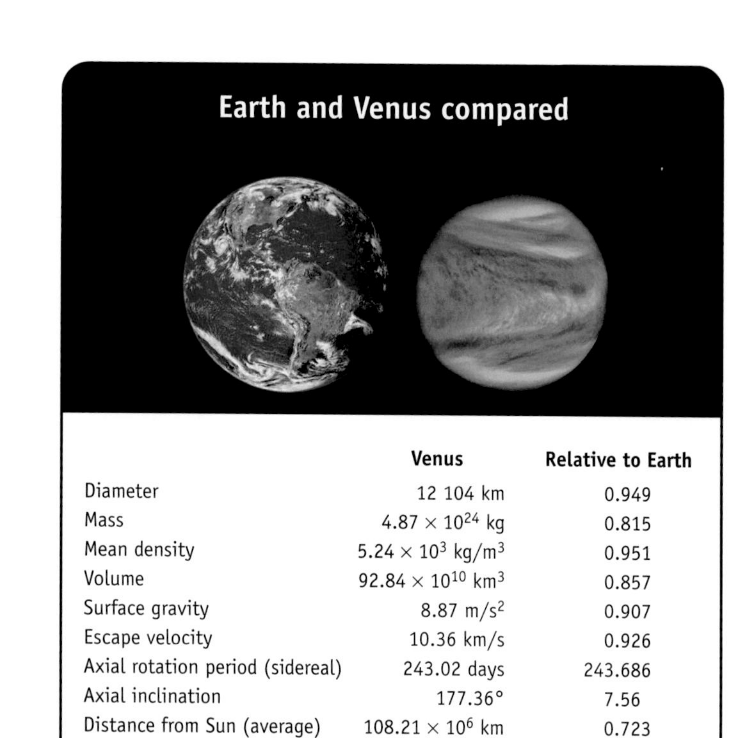

	Venus	Relative to Earth
Diameter	12 104 km	0.949
Mass	4.87×10^{24} kg	0.815
Mean density	5.24×10^{3} kg/m^3	0.951
Volume	92.84×10^{10} km^3	0.857
Surface gravity	8.87 m/s^2	0.907
Escape velocity	10.36 km/s	0.926
Axial rotation period (sidereal)	243.02 days	243.686
Axial inclination	177.36°	7.56
Distance from Sun (average)	108.21×10^{6} km	0.723
Eccentricity of orbit	0.0067	0.401
Orbital period (sidereal)	224.70 days	0.615

▶ **Wrapped up:** Venus is swathed in unbroken clouds, which scud around the planet from east to west every 4 days, 60 times faster than the planet rotates. This photograph was taken by the Pioneer Venus orbiter.

▼ **Clouded out:** Even the mighty Hubble Space Telescope is unable to see anything more than vague markings in the clouds of Venus.

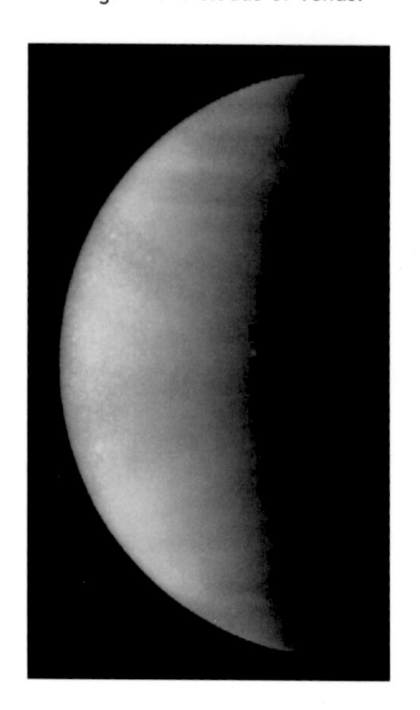

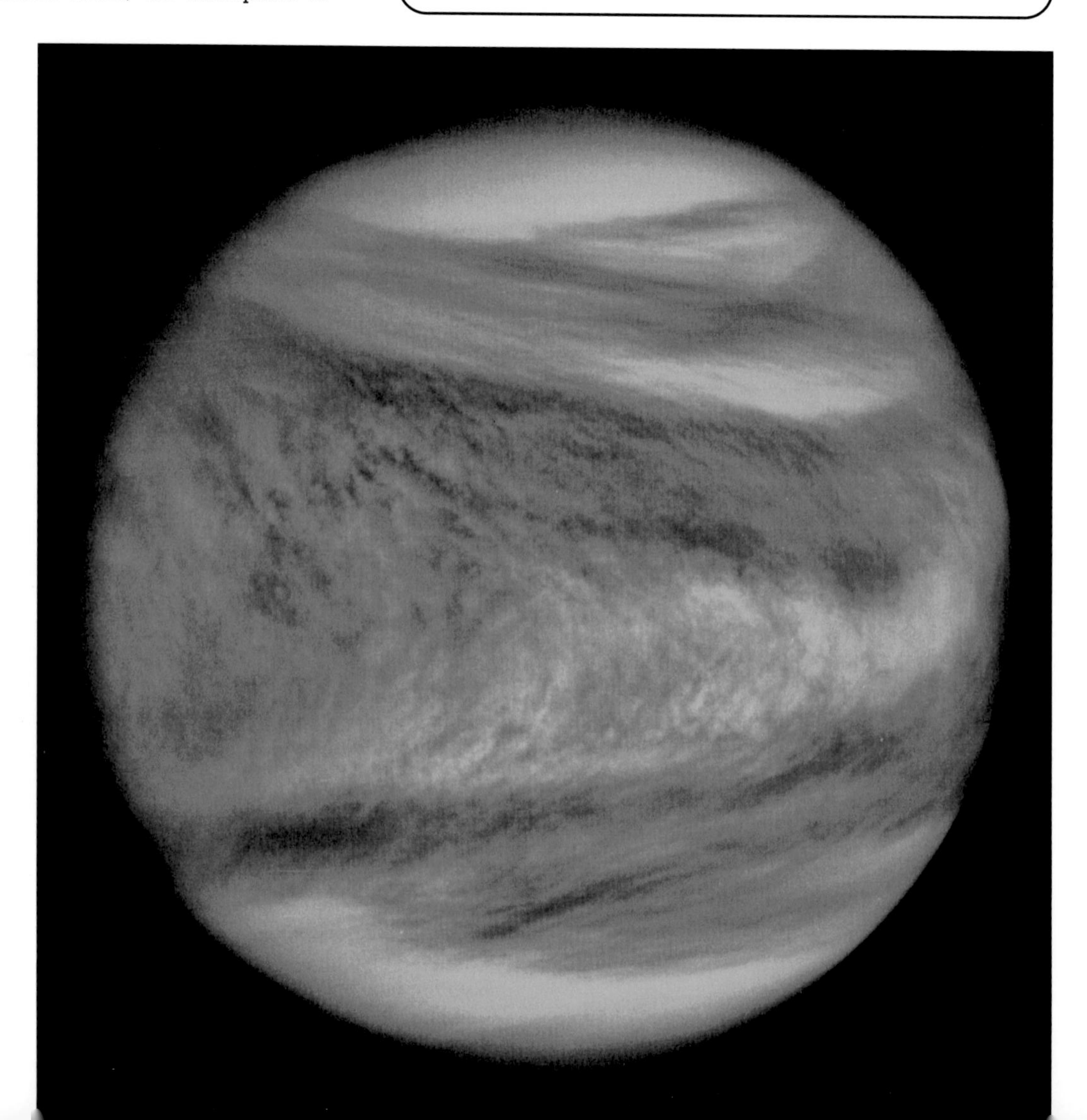

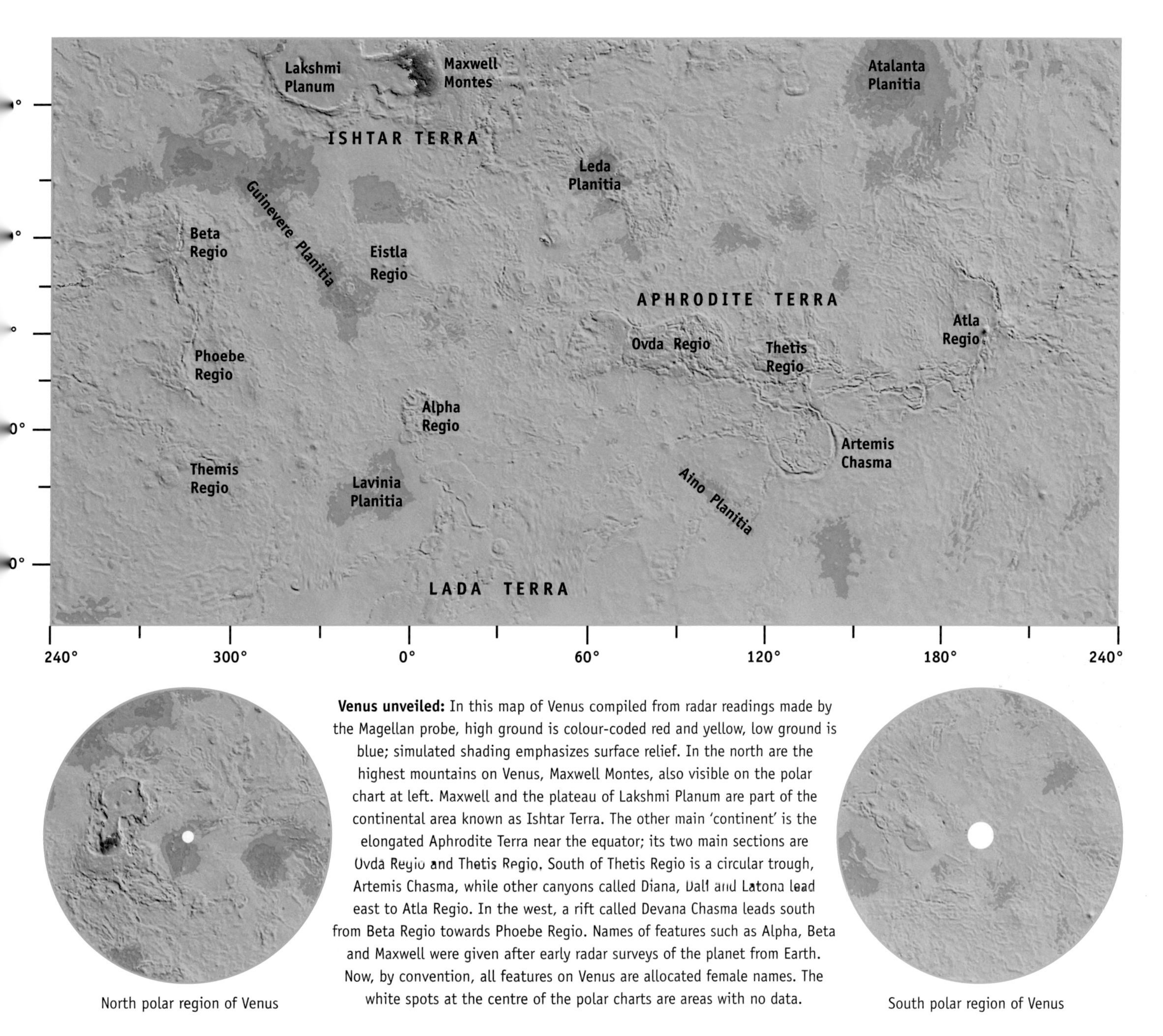

North polar region of Venus

Venus unveiled: In this map of Venus compiled from radar readings made by the Magellan probe, high ground is colour-coded red and yellow, low ground is blue; simulated shading emphasizes surface relief. In the north are the highest mountains on Venus, Maxwell Montes, also visible on the polar chart at left. Maxwell and the plateau of Lakshmi Planum are part of the continental area known as Ishtar Terra. The other main 'continent' is the elongated Aphrodite Terra near the equator; its two main sections are Ovda Regio and Thetis Regio. South of Thetis Regio is a circular trough, Artemis Chasma, while other canyons called Diana, Dali and Latona lead east to Atla Regio. In the west, a rift called Devana Chasma leads south from Beta Regio towards Phoebe Regio. Names of features such as Alpha, Beta and Maxwell were given after early radar surveys of the planet from Earth. Now, by convention, all features on Venus are allocated female names. The white spots at the centre of the polar charts are areas with no data.

South polar region of Venus

Venus presses down at the surface with a force 90 times that of the Earth's atmosphere. Little wonder that the first space probes to descend into this forbidding cauldron were crushed like tin cans. Later probes, better reinforced, reached the surface successfully and found themselves sitting on flat slabs of volcanic rock.

Radar can cut through the all-enveloping clouds of Venus to reveal the planet's surface features. The best maps have come from an American probe called Magellan which orbited the planet in 1990–94. Magellan's radar charted a gently undulating landscape, punctuated by volcanoes and impact craters. There are two main upland areas, analogous to continents on Earth: Ishtar Terra in the north, similar in size to the United States, containing Maxwell Montes, a range of peaks rising higher than Mount Everest; and the elongated Aphrodite Terra near the equator, some 15,000 km (9000 miles) long, covering about the same area as South America.

Venus spins more slowly on its axis than any other planet, once every 243 days, and in a direction opposite to that of the Earth, from east to west. In fact, Venus takes longer to spin on its axis than it does to orbit the Sun, a circumstance that is unique in the Solar System. Perhaps Venus was victim of a major impact early in its history that slowed its rotation. It seems that all three inner planets – Mercury, Venus and Earth – may have suffered a serious collision at some time during or shortly after their formation, but with a different outcome in each case.

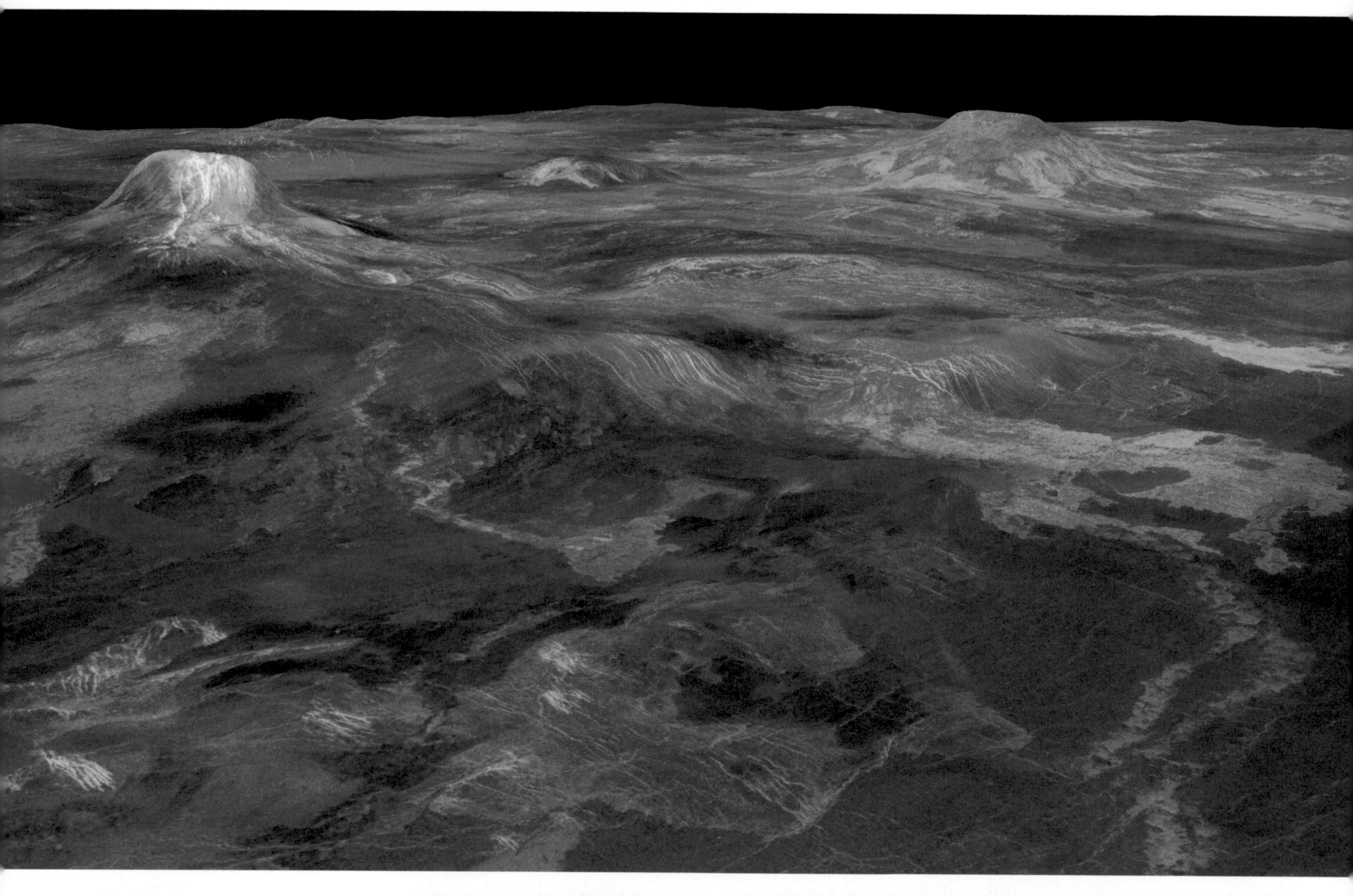

▲ **Overflowing:** Bright, young flows of lava run down the slopes of the volcanoes Gula Mons, upper left, and Sif Mons, upper right, and extend for hundreds of kilometres over the surface of the Eistla Regio highlands. This imaginary view, as would be seen from an altitude of 7.5 km (4.6 miles), was computer-generated from radar data returned by the Magellan space probe. Gula Mons is 3 km (1.8 miles) high, Sif Mons 2 km (1.2 miles) high, and they lie 730 km (450 miles) apart. The heights have been exaggerated 22½ times for clarity and colour has been added to simulate the appearance under the clouds of Venus. Although the lava seen here has long since solidified, some of the volcanoes on Venus may still be active.

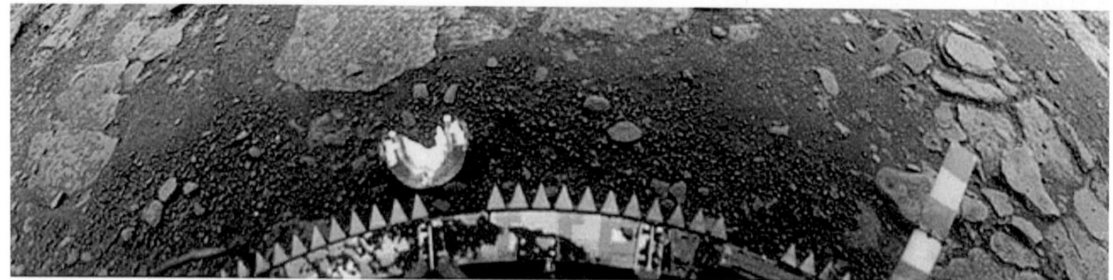

▲ **Colour me yellow:** Under the pall of its clouds, Venus is bathed in a sulphurous glow, captured on this first colour picture from the surface sent by the Russian probe Venera 13 in 1982 (top). The same panorama is also shown with the atmospheric coloration removed, demonstrating that the rocks of Venus are in fact a dull volcanic grey. Part of the Venera lander itself is visible at the bottom of the picture, with a discarded lens cap on the surface at centre left and a colour chart at the right. The landing site was to the east of Phoebe Regio.

◀ **Three strikes:** A trio of large impact craters on Lavinia Planitia, a lowland in the southern hemisphere of Venus, in a view computer-generated from Magellan radar data. The crater in the foreground is Howe, 37 km (23 miles) wide, with Danilova (48 km, 30 miles) wide at upper left and Aglaonice (63 km, 39 miles) wide at upper right. All three have central peaks and are surrounded by aprons of ejecta splashed out by the crater-forming impacts. Large meteorites can penetrate the dense atmosphere of Venus and the surface, but smaller ones burn up. Hence there are no impact craters on Venus smaller than about 3 km (2 miles) diameter. The largest impact crater on Venus is some 280 km in diameter, larger than almost all the craters on the Moon.

▼ **Lava pancakes:** These flat-topped domes on the eastern edge of Alpha Regio are thought to have been formed by the welling up of thick, sticky lava through vents in the surface of Venus. They are about 25 km (15 miles) wide but only 750 m (2500 ft) high; this computer-generated view from Magellan radar data artificially exaggerates their vertical relief.

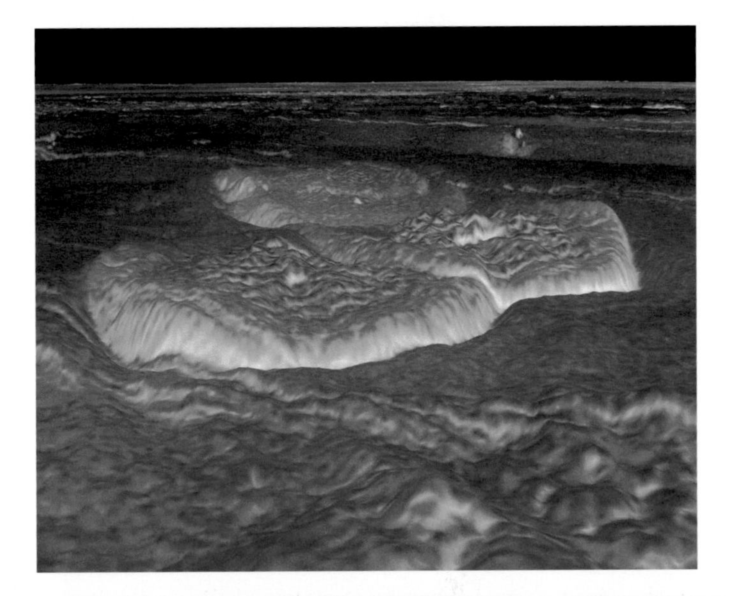

▲▼ **Flatland:** After Maxwell Montes, the highest peak on Venus is Maat Mons, a shield-like volcano 8 km (5 miles) high in the Atla Regio uplands at the eastern end of Aphrodite Terra. Lava flows can be seen spilling down its flanks on the view above. The vertical scale of this view has been exaggerated 22½ times, making the surface appear unrealistically rugged. When the same picture is reproduced with its natural proportions, below, the sides of Maat Mons are transformed into gentle slopes typical of shield volcanoes on Earth, emphasizing how flat Venus really is.

Mars, the red world

The first planet beyond Earth on which humans will set foot is Mars. It is a world half the size of our own with a similar length of day, an atmosphere frosted with cirrus clouds, and icy polar caps that shrink in summer and expand in winter. Above all, its sandy red surface has long seemed the most likely place in the Solar System to find other life forms.

A century ago an American astronomer, Percival Lowell, created a sensation when he claimed to see long, straight markings criss-crossing the deserts of Mars. He conjectured that these were irrigation canals dug by Martian inhabitants struggling against drought, a romantic speculation that inspired a genre of science fiction stories culminating with H. G. Wells's *War of the Worlds*. Other astronomers dismissed Lowell's canals as tricks of the eye compounded by an overactive imagination, but acknowledged that darker areas of the surface, for example the prominent Syrtis Major, might be tracts of lowly vegetation.

Those hopes were dashed when space probes reported on the harsh reality of Mars. The planet's atmosphere is as thin as at an altitude of some 32 km (20 miles) above the Earth, and is composed primarily of carbon dioxide with scarcely any free oxygen. On Earth, oxygen is released by photosynthesis in plants so its deficiency on Mars argues strongly against the existence of vegetation there. This conclusion was reinforced in 1976 by two American landers called Viking which scooped up samples of the soil and analysed them in an unsuccessful search for Martian bugs. Such negative results should hardly be surprising in view of what else we now know of conditions on the planet – with only scant atmosphere to protect it, the surface is sterilized by ultraviolet light from the Sun and temperatures rarely rise above the freezing point of water even on a summer afternoon. Space suits will be essential attire for Martian explorers.

However, the climate of Mars has not always been so hostile. Photographs taken by probes orbiting Mars show apparent dried-up river beds – evidence that much of the planet was awash with liquid water billions of years ago, albeit temporarily. Such outpourings would have accompanied the eruption of volcanoes which pumped out gases to create a denser, warmer atmosphere. During those more clement times, it is possible that life gained a foothold on the planet. Sheltered inside rocks or underneath the surface, some resilient organisms from that period may still lie dormant.

In 1996 a group of NASA scientists reported traces of ancient microbial life in a meteorite that is thought to have come from the surface of Mars, but their conclusions remain controversial. Future space probes will burrow into the soil and look under rocks for organisms. Even so, we may not know for sure whether Mars harbours life until probes bring rock samples back to Earth.

Earth and Mars compared

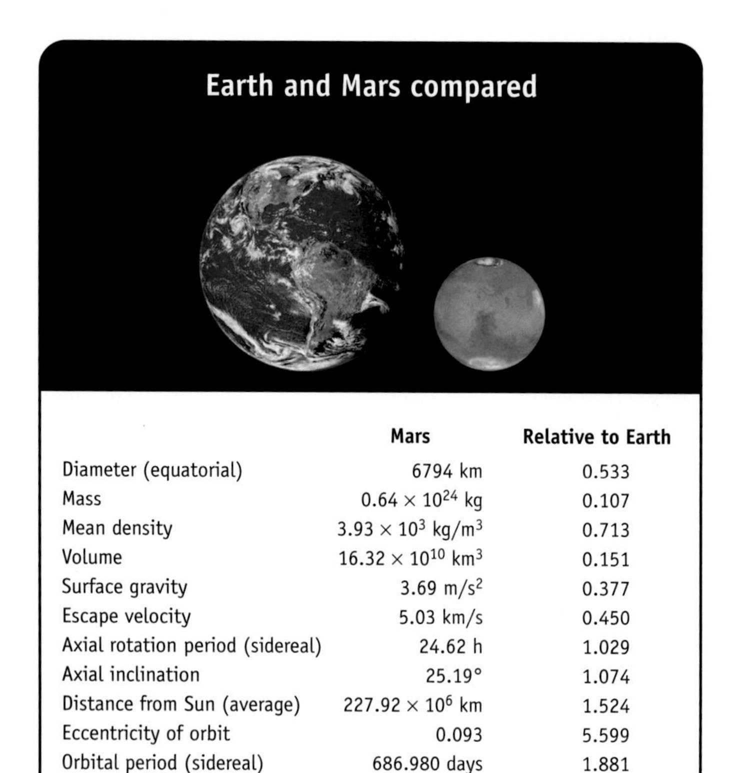

	Mars	Relative to Earth
Diameter (equatorial)	6794 km	0.533
Mass	0.64×10^{24} kg	0.107
Mean density	3.93×10^{3} kg/m^3	0.713
Volume	16.32×10^{10} km^3	0.151
Surface gravity	3.69 m/s^2	0.377
Escape velocity	5.03 km/s	0.450
Axial rotation period (sidereal)	24.62 h	1.029
Axial inclination	25.19°	1.074
Distance from Sun (average)	227.92×10^{6} km	1.524
Eccentricity of orbit	0.093	5.599
Orbital period (sidereal)	686.980 days	1.881

THE MOONS OF MARS – CAPTURED ASTEROIDS?

Mars has two small, misshapen moons: Phobos, pictured below, 27 km (17 miles) across at its widest, and Deimos, right, 15 km (9 miles) wide. Phobos sports a deep impact crater at one end with radiating fractures, while Deimos has a smoother surface. Both are thought to be escapees from the asteroid belt that were captured by the gravity of Mars.

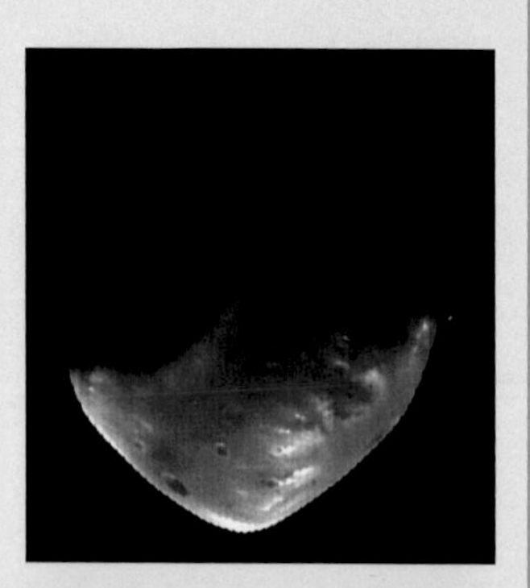

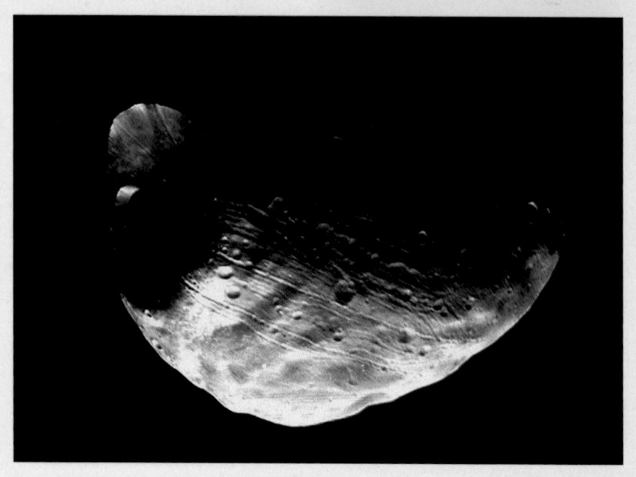

HUBBLE'S VIEW OF MARS

Dark markings on the ruddy face of Mars change in size and shape with the seasons. Once surmised to be areas of vegetation, these are now known to be expanses of darker rock periodically covered and uncovered by sand and dust as it is blown around by seasonal winds. Most emblematic of the surface markings of Mars is the tongue-shaped Syrtis Major, first seen telescopically in the 17th century by the Dutch astronomer Christiaan Huygens. Syrtis Major is central in this image from the Hubble Space Telescope, taken in 1999 during one of Mars's biennial approaches to Earth. Mars was then 87 million km (54 million miles) away, close enough for Hubble to resolve individual impact craters on its surface. South of Syrtis Major is a huge lowland basin called Hellas, the scar of an asteroid strike over 2000 km (1200 miles) wide, larger than Mare Imbrium on the Moon; here it is whitened by surface frosts and cirrus clouds. At the upper right, late afternoon clouds have formed around Elysium, a volcanic upland area. This view was taken during northern hemisphere summer on Mars. The north polar cap is tilted Earthwards so latitudes below 60° south cannot be seen.

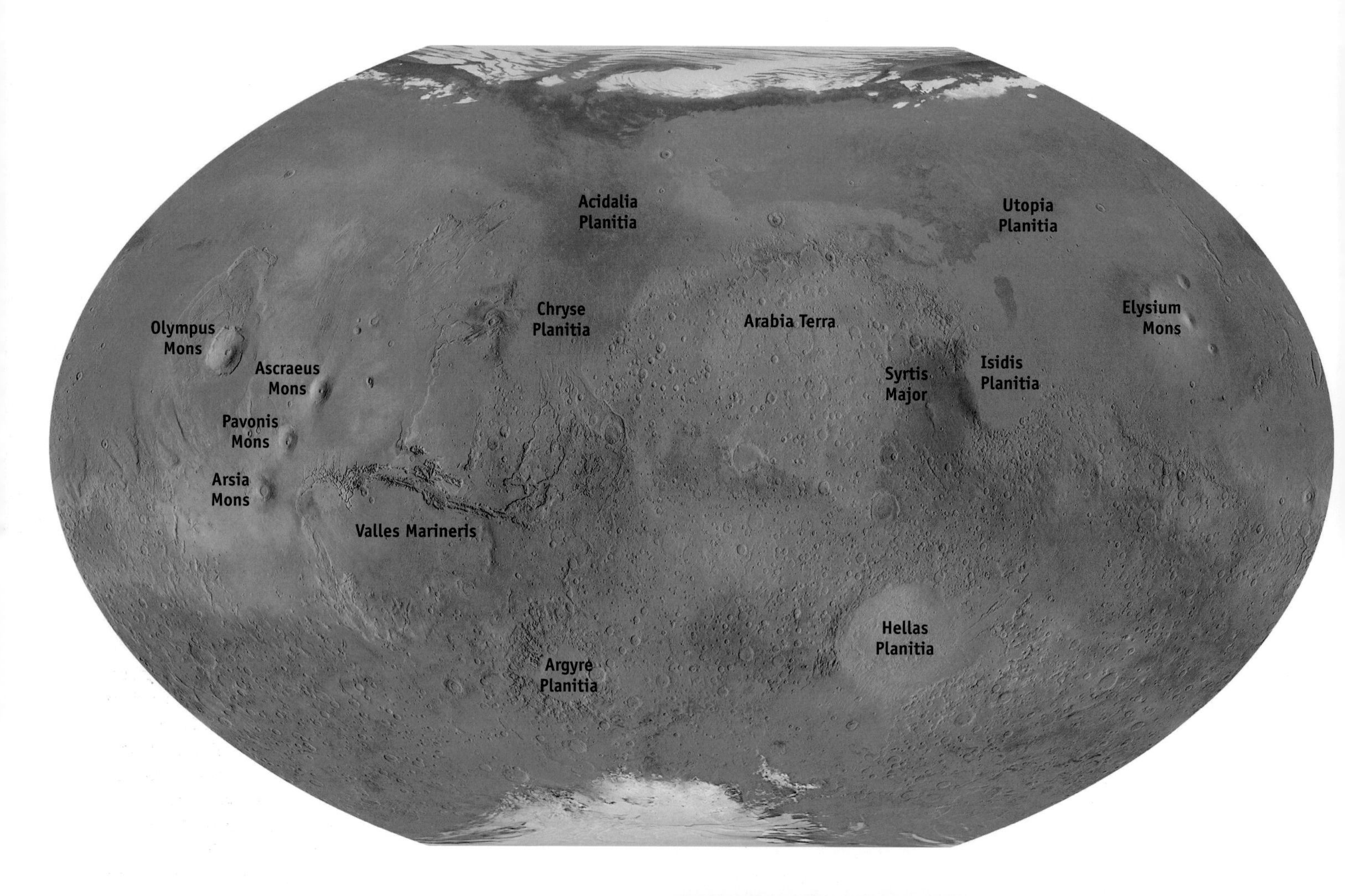

▲ **Global picture:** This map of the entire globe of Mars was assembled from about a thousand wide-angle images and over 200 million laser altimeter measurements taken by NASA's Mars Global Surveyor spacecraft. The altimetry accentuates surface relief and the image data provides realistic colour; the image projection is of the type known as Winkel–Tripel. Major features are labelled, notably the quartet of large volcanoes at left with the extensive Valles Marineris rift valley to the east of them, and the Hellas and Argyre impact basins in the southern hemisphere. The Viking 1 and Viking 2 landers touched down on two lowland plains, Chryse and Utopia, in 1976. The Mars Pathfinder lander also came down in Chryse in 1997. Isidis Planitia, east of Syrtis Major, is the target for the European Beagle 2 lander in December of 2003 which will continue the hunt for elusive Martian life. Compare this view with those on pages 40 and 41 which are colour-coded to emphasize contours.

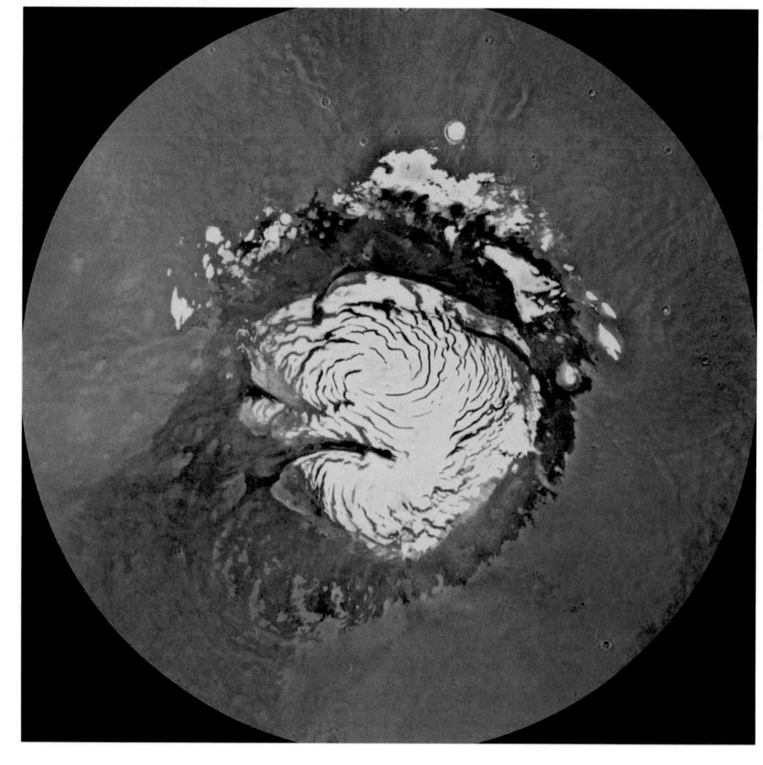

▶ **Frozen north:** Mars has polar caps with a dual composition of ordinary water ice and frozen carbon dioxide ('dry ice'). Carbon dioxide freezes out of the atmosphere onto the polar cap in the winter hemisphere, but returns to the atmosphere in warmer summer while the water in the caps remains frozen. This composite of images from the Viking orbiter probes depicts the permanent water-ice component of the north polar cap, about 1000 km (600 miles) across. Around the cap is a dark collar consisting of sand dunes.

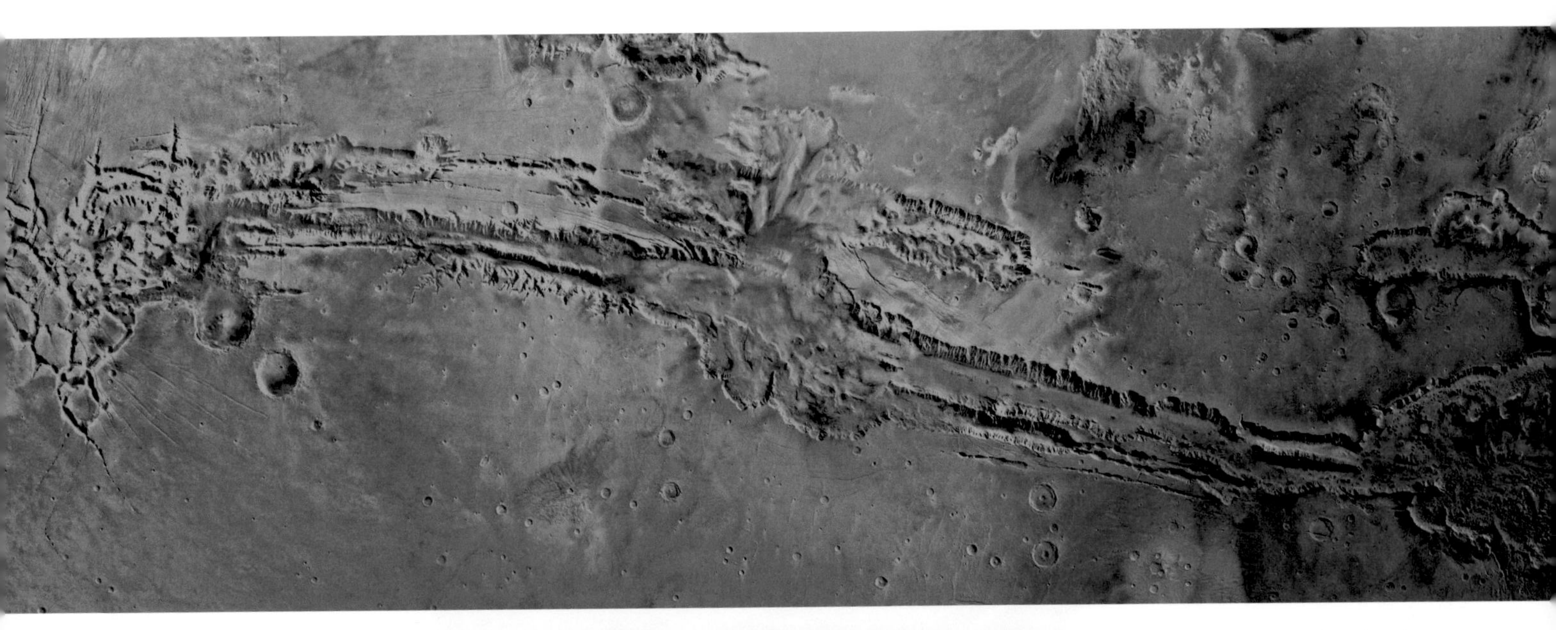

▲ **Rift valley:** Long enough to stretch across the United States from Los Angeles to New York, Valles Marineris is the greatest chasm on any planet in the Solar System. It starts in the west with the gridded triangular region known as Noctis Labyrinthus and extends for more than 4000 km (2500 miles), opening out into the northern Chryse lowlands where Viking 1 and Mars Pathfinder landed. The whole complex is thought to have originated by faulting in the Martian crust and subsequently was eroded by winds and water. This image is a composite of views from the Viking orbiter probes.

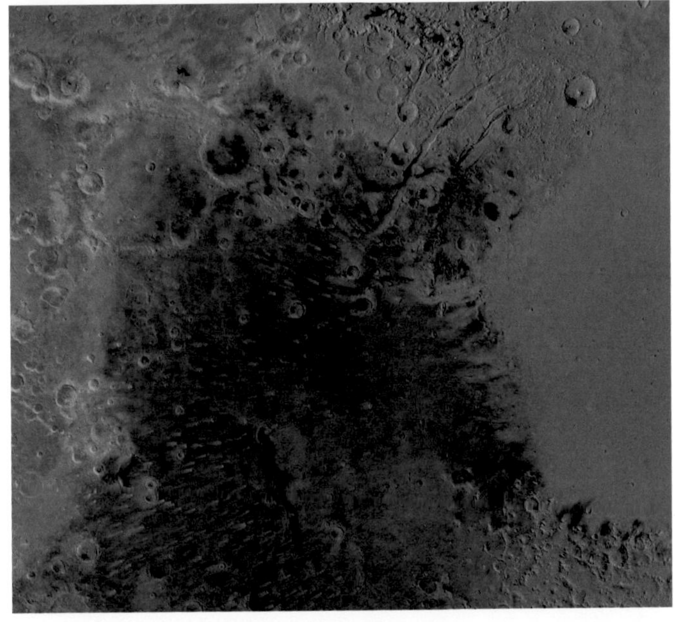

◀ **Dark secret:** Syrtis Major, the most distinctive surface marking on Mars as seen through a telescope, turns out to be an otherwise unremarkable highland region of dark dust and lava flows sloping down to the brighter surface of the Isidis basin to the east. Numerous impact craters and streaks of wind-blown dust are visible in this image of Syrtis Major, assembled from pictures taken by the Viking orbiter probes.

▶ **King cone:** Three times the height of Mount Everest and covering a similar area to Arizona, Olympus Mons is the largest volcano in the Solar System. This view from the west has been computer-generated from height readings taken by Mars Global Surveyor. The deposits in the foreground are landslips, not lava flows, but there are signs that this volcano has erupted within the past 100 million years. The vertical scale is exaggerated ten times; in reality it would be easy to stroll up the slopes.

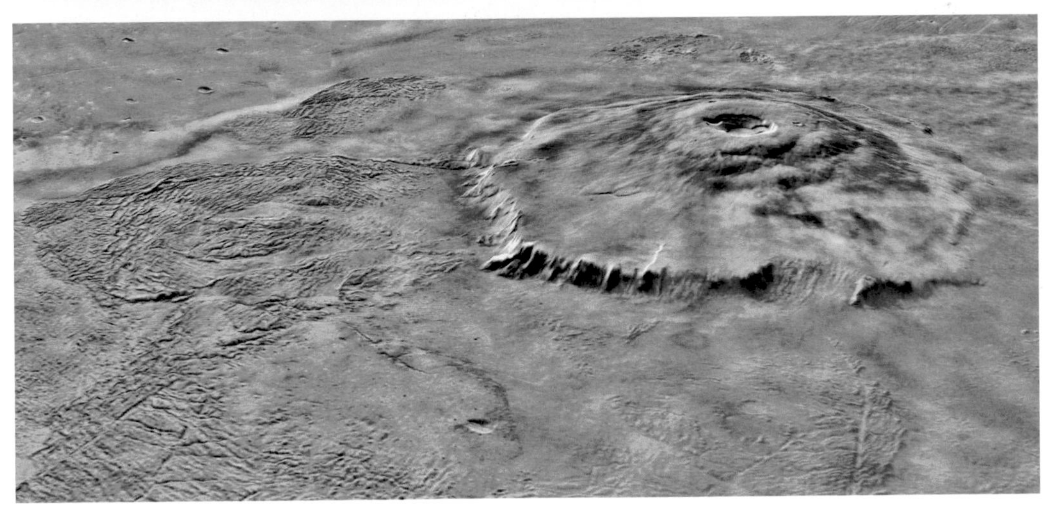

▲ **Desert view:** Rocks and sand cover the surface of the Chryse lowland plain in this 360° panorama taken by the Mars Pathfinder spacecraft in 1997. Part of the lander is visible in the foreground, while in the middle distance the Sojourner rover is examining a boulder dubbed Yogi. Although Mars is traditionally known as the red planet, its true colour is more of a golden brown, due to iron oxide – most familiar on Earth as rust. Fine dust suspended in the atmosphere imparts the same colour to the sky. Many of the rocks in this picture are thought to have been washed here by catastrophic floods billions of years ago.

◀ **Rolling on:** Sojourner, a six-wheeled miniature rover carried by the Mars Pathfinder lander, casts long shadows in the evening sunshine after rolling down the ramp at lower left onto the surface of Mars in July 1997. Sojourner, which operated on the surface for nearly three months, carried a device to analyse the composition of rocks but did not look for life.

▼ **Sundown:** Evening on Mars, seen from the planet's surface by Mars Pathfinder. Silhouetted at the left against the dying rays of the Sun are two low hills known as Twin Peaks, about 1 km away and visible on the panorama above.

◀▼▶ **Outstanding:** These rocks will appear three-dimensional when viewed through standard 3D glasses, with a red filter over the left eye and a blue filter over the right. Mars Pathfinder scientists gave the rocks catchy names. Clockwise from right they are Grommit, Barnacle Bill, Wedge and Yogi (with the smaller Boo Boo in the background). Yogi is the largest, about a metre across, and is seen in the panorama above being studied by Sojourner.

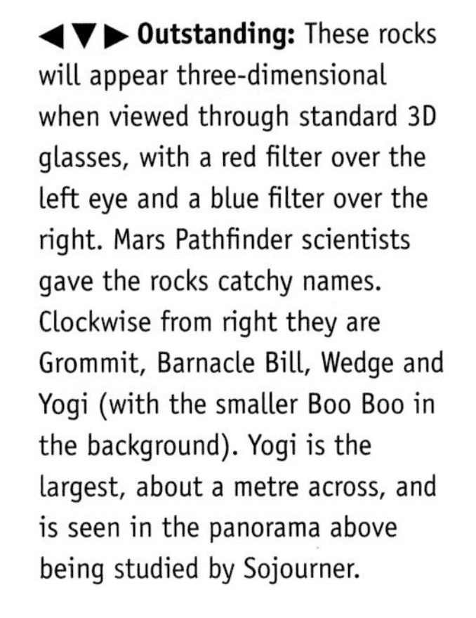

◀ **Deep impact:** The highs and lows of Mars are displayed graphically in these relief views of two opposite hemispheres. Colour-coded altitudes emphasize that Mars is a planet of two halves: a smooth, low northern hemisphere and cratered uplands in the south. Puncturing the southern highlands is the largest impact basin on Mars, Hellas Planitia, shown as dark blue; it occupies an area almost half that of the United States and is deep enough to swallow Mount Everest. Ejecta from the impact is spread over the surrounding terrain for thousands of kilometres. Hellas is carpeted with dust, and dust storms frequently occur within it. Above left of Hellas is a prominent impact crater, Huygens, 450 km (280 miles) across. Farther north is the Syrtis Major region, bounded on its east by the horseshoe-shaped lowland of Isidis Planitia. At the upper right is the volcanic mountain Elysium Mons. Yellow corresponds to the Martian equivalent of sea level; green and blue are lower, orange and red higher, white the highest of all. Artificial illumination imparts three-dimensional shading.

WATER ON MARS

Although there is no liquid water on the surface of Mars today, there are unmistakable signs that it flowed freely there in the past. Some flows date back to early in the planet's history when its climate is thought to have been milder. Others are more recent, and are more difficult to account for. Proposed explanations for the more recent flows include the melting of subsurface permafrost by volcanic heating, or leakage of liquid water from beneath the permafrost either gradually or in a sudden gush. The three apparent dried-up rivers shown in this oblique view from Mars Global Surveyor lie to the north-east of the Hellas basin, and can be located on the relief globe above. From the top they are Dao Vallis, Niger Vallis (which joins Dao about halfway) and Harmakhis Vallis. The valleys, many hundreds of kilometres long, are roughly 1 km (0.6 mile) deep and range in width from about 40 km (25 miles) to 8 km (5 miles). All three become narrower as they drain downhill into Hellas. There is currently no way of telling when these valleys formed.

▶ **High spots:** The opposite face of Mars from the Hellas basin is dominated by a huge upland called Tharsis, which straddles the Martian equator. If transplanted to Earth, it would stretch down the west coast of the United States from the border of Canada to Mexico. This enormous swelling is punctuated by three towering volcanic peaks, Ascraeus Mons, Pavonis Mons and Arsia Mons. To the northwest of Tharsis is the largest volcano of all, Olympus Mons, cresting more than 21 km (13 miles) above the local equivalent of sea level. To the north is a more subdued volcanic rise, Alba Patera, scarcely noticeable on the map on page 36 because of its shallow slopes. Clouds are common over the Tharsis volcanoes in mid-afternoon, often appearing to form a letter W as viewed from Earth. Eastern Tharsis is sliced through by the Valles Marineris rift. From the peaks of the volcanoes to the depths of Hellas (facing page), the full range of elevations on Mars is about 30 km (19 miles), one and a half times greater than on Earth and more than on any other planet.

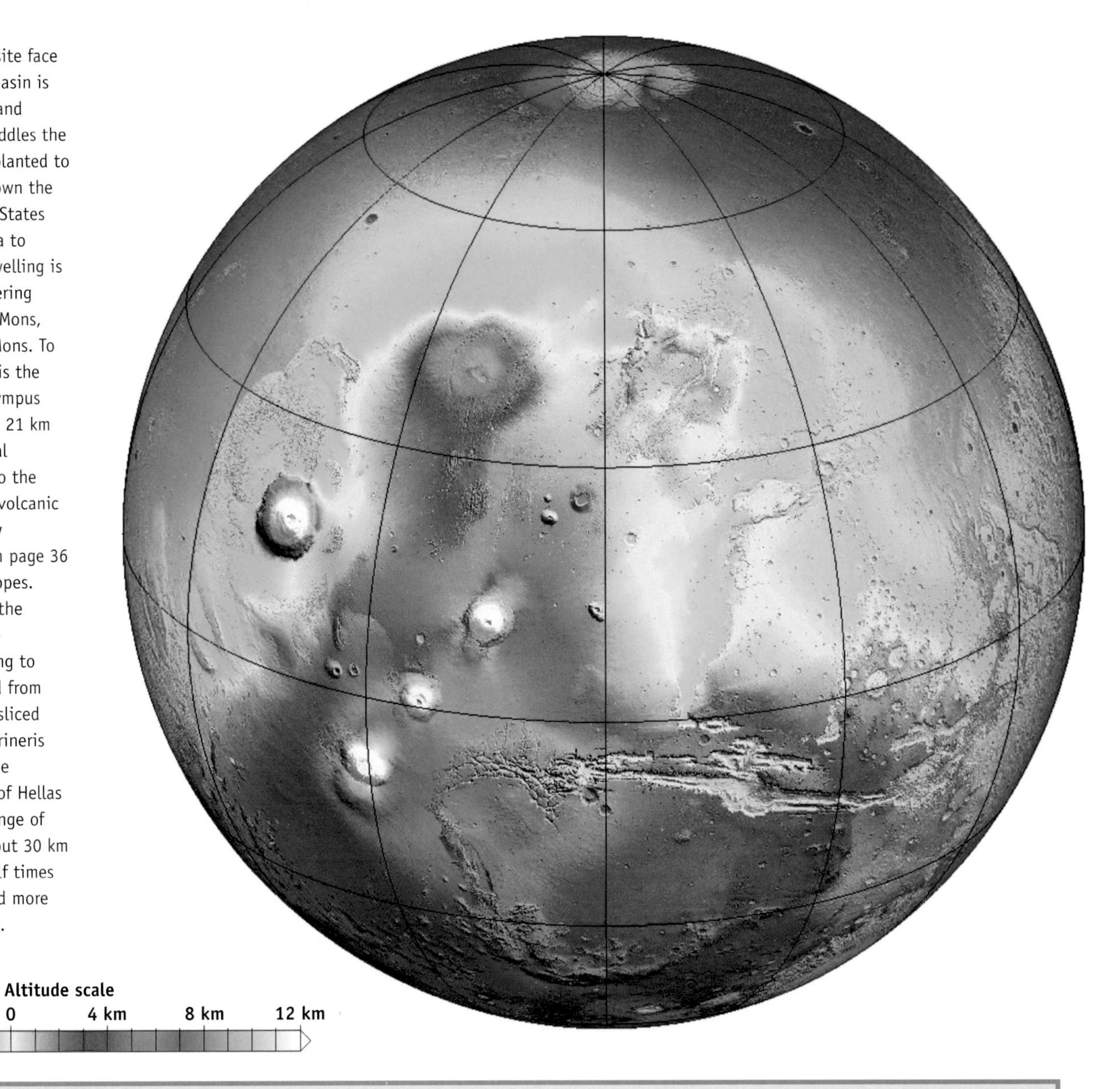

Altitude scale

−8 km | −4 km | 0 | 4 km | 8 km | 12 km

THE 'FACE' ON MARS

In 1976, when Viking 1 photographed a rocky mesa in Cydonia, an area on the southeastern border of Acidalia Planitia, the lighting effects gave it the appearance of a pharaoh-like face. NASA published the picture as a curiosity, but among certain sections of the public the opinion grew that the sculptured outcrop was in some way artificial. In 2001, Mars Global Surveyor photographed the 'face' again, in far greater detail. Here, the views from Viking and Mars Global Surveyor are shown to the same scale for comparison.

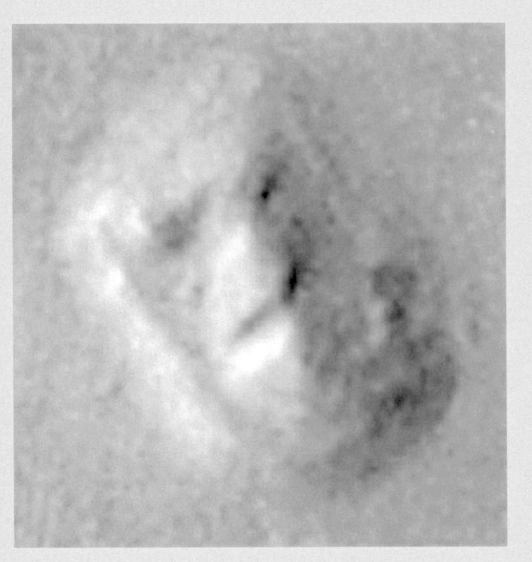

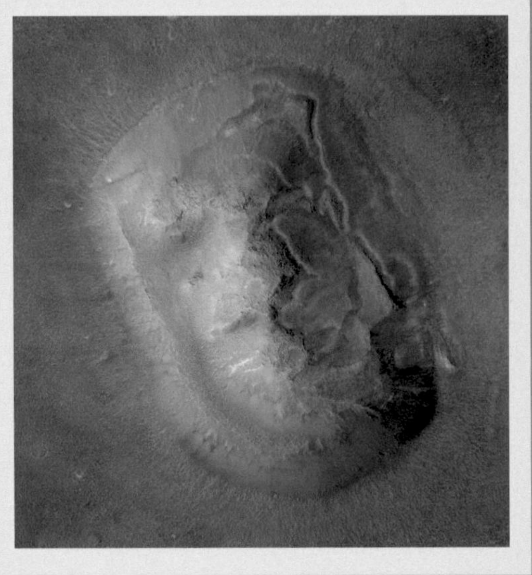

Asteroids, interplanetary debris

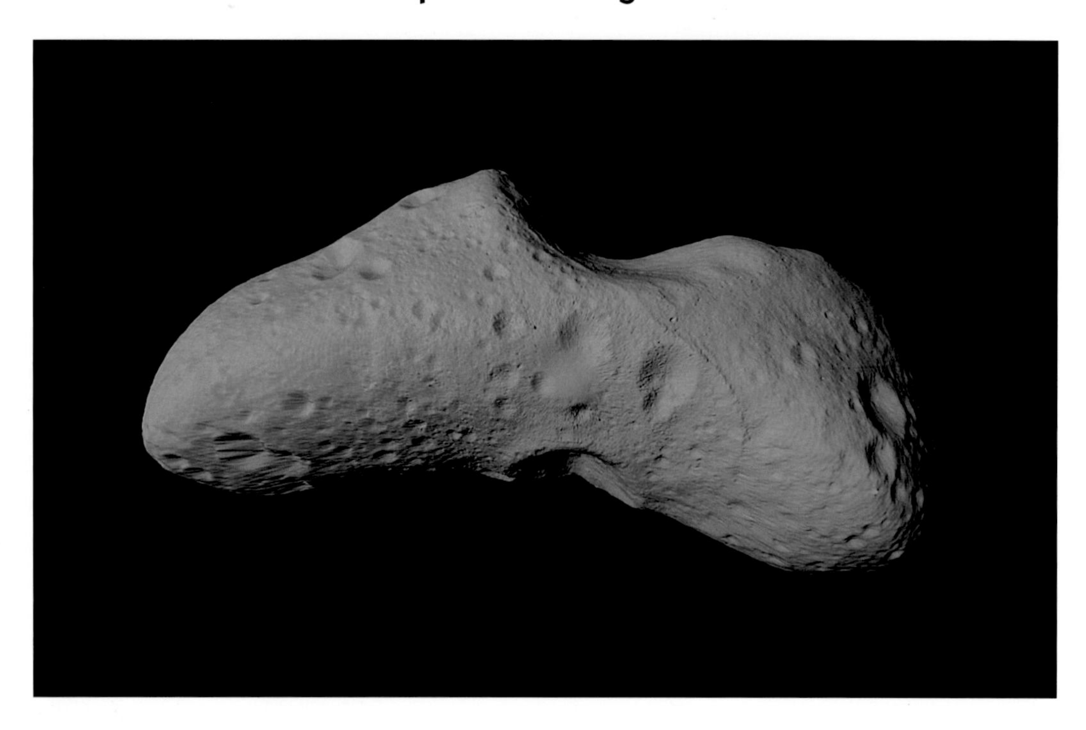

◀ **Making eyes at Eros:** The highly elongated asteroid Eros, 33 km (21 miles) from end to end, exhibits a butterscotch hue in this true-colour representation created from photographs and laser rangefinder readings returned by the NEAR Shoemaker probe which went into orbit around it in February 2000 (NEAR stands for Near-Earth Asteroid Rendezvous, and Eugene Shoemaker was an American asteroid expert who died in 1997). Eros is one of a small group of asteroids that can come close to Earth, but in this case not close enough to pose a collision threat. The large saddle-shaped indentation at the top is called Himeros, and the impact crater on the opposite side is Psyche, seen in more detail below.

Beyond the orbit of Mars, millions of flying chunks of rock and metal course around the Sun along a broad ring road known as the asteroid belt. These bodies are termed 'asteroids' from their starlike appearance through telescopes (Greek *aster*: star); they are also known as minor planets. Ceres, the largest, is 940 km (580 miles) across and would fit comfortably into the Gulf of Mexico, but most range in size from pebbles to mountains. Collected together, the asteroids would form a single body far smaller than the Moon.

The Earth and other major planets are thought to have grown from the accumulation of bodies like the asteroids some 4.5 billion years ago. The asteroids we see today are remnants from that process. Originally the asteroid belt probably included several larger members. Mutual collisions have since ground these into smaller, irregular fragments with a mixture of compositions. Their outer layers gave rise to families of rocky asteroids, whereas asteroids composed of iron and nickel came from their cores.

▼ **Craters on Eros:** Psyche, the largest crater on asteroid Eros, is 5.3 km (3.3 miles) wide. Dust and rocks are seen within it in this view from NEAR Shoemaker. Smaller craters on its rim create a paw-like appearance.

▼ **Rocky Eros:** A close-up of the eroded, dusty surface of Eros taken by NEAR Shoemaker. The large rock below and to the right of centre is about the size of a house, while the smallest rocks visible are the size of humans.

◀ **Travelling companion:** Ida, a member of the asteroid belt 55 km (34 miles) long, has a small moon, Dactyl, about 1.6 km (1 mile) wide, seen to the right in this picture taken in 1993 by the Galileo space probe. Dactyl lies about 100 km (60 miles) from Ida.

▼ **Asteroid and moons:** Asteroid Gaspra, top centre, compared with the two moons of Mars, Deimos, lower left, and Phobos. All three are shown to the same scale. Phobos and Deimos are thought to be captured asteroids and, like Gaspra, have irregular shapes that are the legacy of collisions.

Our first close-up look at a member of the asteroid belt came in 1991 when NASA's Galileo space probe, on its way to Jupiter, flew past Gaspra, showing it to be an angular, cratered rock measuring about 18 km (11 miles) at its longest (see right). Two years later Galileo flew past a larger asteroid, Ida, which turned out to have a satellite (pictured above). Since then, other asteroids have been found to have moons or to consist of two similarly sized fragments, a clear consequence of collisions.

THE ASTEROID THAT DISPATCHED THE DINOSAURS

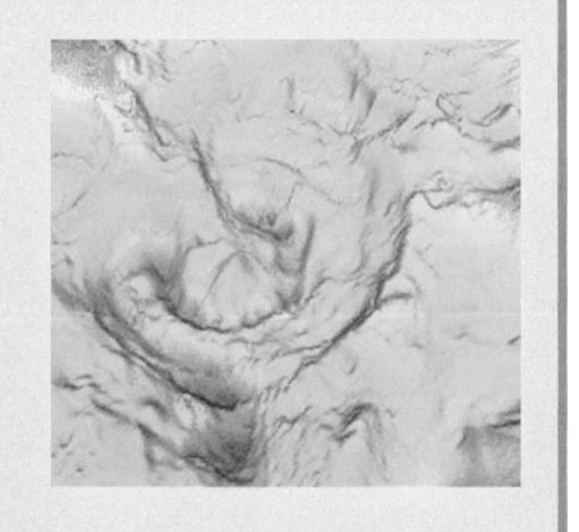

Buried under 1 km of limestone sediments, this crater (right) is thought to mark the spot where an asteroid hit the Earth 65 million years ago, scattering a dense blanket of dust into the atmosphere that altered the climate and extinguished not only the dinosaurs but over two-thirds of all life on Earth. Centred near the city of Mérida in the Yucatán peninsula of Mexico, the 180-km (112-mile) diameter crater was formed by an asteroid about 10 km (6 miles) wide. Because it is covered by sediments laid down since its formation, the crater is not visible at the surface. Its existence and age were established only in the 1990s as a result of small variations in the local gravitational field and studies of rocks from drillings by oil prospectors. This false-colour map shows the crater's outline as it appears in gravity data. The crater is named Chicxulub, after the location of the first well drilled in the region.

A small proportion of asteroids have been exiled from the main belt by the gravitational dictate of dominant Jupiter, and now wander blindly across the orbits of the inner planets. Several telescopic surveys keep nightly watch for asteroids that might be on a collision course with us, mindful of the consequences to the dinosaurs when an asteroid squarely struck the Earth 65 million years ago (see box). One famous object that strays beyond the asteroid belt, Eros, was surveyed in detail by NASA's NEAR Shoemaker probe, which orbited it for a year before ending its mission by gently descending to the asteroid's rocky surface.

Small chips from asteroids regularly fall to Earth as meteorites. Some hit with sufficient force to blast out craters such as Meteor Crater in Arizona but others, which land more gently, survive and can be collected. Hence the meteorites in our museums are samples of asteroids. Dating them by measuring the proportions of radioactive elements they contain tells us the age of the Solar System.

Jupiter, the giant planet

If there could be said to be a dominant male among the planetary pride then Jupiter is it. Eleven times the diameter of the Earth and over 300 times as massive, Jupiter, named after the king of the Roman gods, is undisputed king of the planets.

According to modern theories of Solar System formation, Jupiter was the first planet to take shape from the cloud of gas and dust that surrounded the young Sun. Fast-growing Jupiter grabbed all the gas within its gravitational reach in the few million years before the cloud dispersed into space. As a result, Jupiter is composed predominantly of hydrogen and helium, also the main components of the Sun, in marked contrast to the four rocky inner planets. This gaseous composition is shared by the more-distant Saturn and, to a lesser extent, Uranus and Neptune.

Even a modest back-garden telescope reveals that the face of Jupiter is marbled by multicoloured clouds, which are drawn out into bands running parallel to the equator by its rapid rotation – once in under 10 hours, the fastest of all the planets. Cloud features track across Jupiter's disk as it spins, but almost all are transitory. Some oval-shaped white storm clouds have been followed for over half a century, but the nearest thing that Jupiter has to a permanent feature is the Great Red Spot (see below).

Conditions within Jupiter's clouds were experienced first-hand in December 1995 by a small probe released by the Galileo spacecraft. Buffeted by winds of up to 720 km/h (500 mile/h), the descending probe transmitted for an hour until it was destroyed by the combined effects of a temperature of 150°C and a pressure 22 times that at the Earth's surface, nearly 200 km (125 miles) below the visible cloud tops. Deeper still, pressures and temperatures increase inexorably. At a depth of about 10,000 km (6000 miles), conditions become so extreme that hydrogen atoms are broken down into a dense, ionized liquid that conducts electricity like a metal, powering the planet's extensive magnetic field. At its centre, Jupiter may have a solid core about 10 times the mass of the Earth.

Earth and Jupiter compared

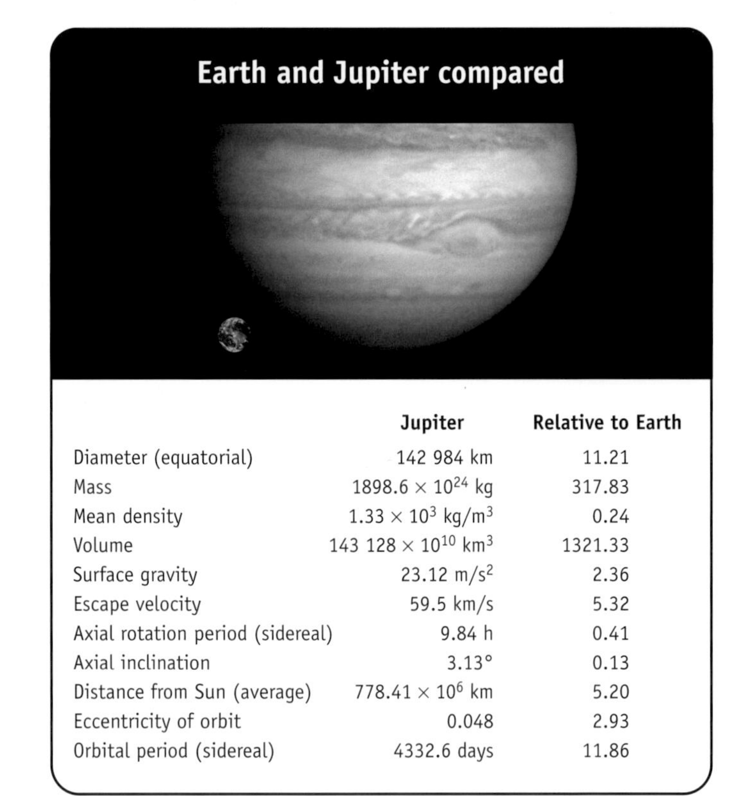

	Jupiter	Relative to Earth
Diameter (equatorial)	142 984 km	11.21
Mass	1898.6×10^{24} kg	317.83
Mean density	1.33×10^{3} kg/m^{3}	0.24
Volume	$143\ 128 \times 10^{10}$ km^{3}	1321.33
Surface gravity	23.12 m/s^{2}	2.36
Escape velocity	59.5 km/s	5.32
Axial rotation period (sidereal)	9.84 h	0.41
Axial inclination	3.13°	0.13
Distance from Sun (average)	778.41×10^{6} km	5.20
Eccentricity of orbit	0.048	2.93
Orbital period (sidereal)	4332.6 days	11.86

THE RED EYE OF JUPITER

First seen through telescopes in 1831, the Great Red Spot is a swirling storm cloud in Jupiter's southern hemisphere, twice the size of the Earth. Spinning anticlockwise once every week, it is a high-pressure region (an *anticyclone*) caused by a rising column of gas from Jupiter's warm interior. Its top is among the highest points in Jupiter's clouds. The colour of the Great Red Spot is thought to be due to traces of phosphine or possibly sulphur in the planet's atmosphere. This picture was taken by the Voyager 1 space probe in 1979. To the south of the Great Red Spot is one of the white ovals that formed in 1940. Like the Great Red Spot, these white ovals are also anticyclonic storm clouds.

CASSINI VIEWS JUPITER

Deep banks of curdling cloud wrap around Jupiter. Lighter-coloured zones, where gas from Jupiter's warm interior rises and condenses into high-altitude clouds, alternate with darker avenues where the gas descends again. The outline of Jupiter is noticeably squat, its polar diameter being nearly 10,000 km (6000 miles) less than at the equator, a combined result of its high-speed rotation and the fact that it is not a solid body. The white clouds are composed of frozen ammonia, at a temperature of about –145°C. Other colours are due to a mixture of various chemicals, the exact compositions of which remain uncertain, carried on updraughts from beneath the clouds. Red clouds are the highest and hence coldest of all whereas blue regions are the deepest and warmest, covered by high haze. Immense bolts of lightning flash among the clouds and have been photographed on the planet's night side by space probes. This view of Jupiter is a mosaic of four images taken by NASA's Cassini spacecraft in December 2000 as it flew past the planet on the way to its main target, Saturn. The dark spot at left is the shadow of Jupiter's moon Europa.

JUPITER'S MOONS – AND RINGS

As befits such a regal planet, Jupiter is attended by an extensive retinue of moons. The four largest – Io, Europa, Ganymede and Callisto – are easily visible through binoculars from Earth, changing position from night to night as they orbit the planet. Known collectively as the Galilean satellites, because they were discovered in 1610 by the Italian scientist Galileo Galilei, the four are shown here to scale in photographs taken by the appropriately named Galileo space probe, with our own Moon for comparison. Io, similar in size to our Moon, is the most volcanically active body in the Solar System while Europa, smallest of the quartet, has an icy shell that may overlie a deep ocean. Ganymede, the largest moon in the Solar System (larger even than the planet Mercury), has a two-tone surface. Callisto, larger than our Moon, is dark and heavily cratered. In all, at least five dozen moons orbit Jupiter but the full complement is uncertain since additional tiny members are still being discovered. As well as its moons, Jupiter is encircled by a broad, faint ring of dust, discovered by the twin Voyager space probes when they flew past the planet in 1979. The rings lie closer to Jupiter than the innermost Galilean satellite, Io.

◀ **Io:** Yellow and orange sulphur deposits cover the surface of Io, erupted through the moon's many active volcanoes. Io is caught in a gravitational tug of war between Jupiter and the other Galilean satellites which repeatedly squeezes it, releasing frictional heat which melts its interior. This image, assembled from photographs taken by the Galileo space probe, shows Io in natural colour. Dark spots dotted over the surface are volcanic vents – more than 80 have been identified.

▼ **Outstanding:** A volcanic eruption sprays a fountain of dust and sulphur dioxide gas to a height of 140 km (87 miles) above the surface of Io, as spied by the Galileo space probe in 1997.

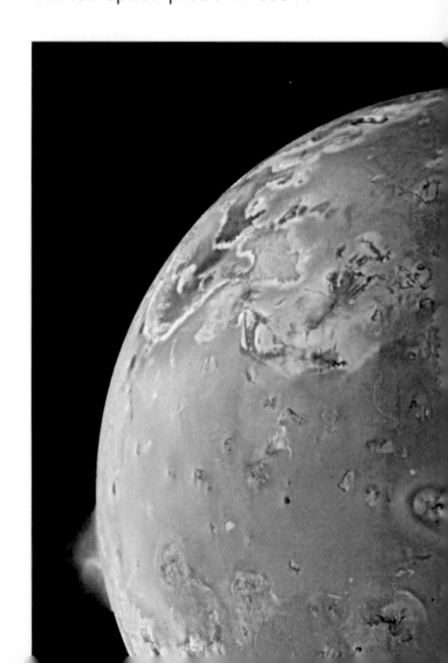

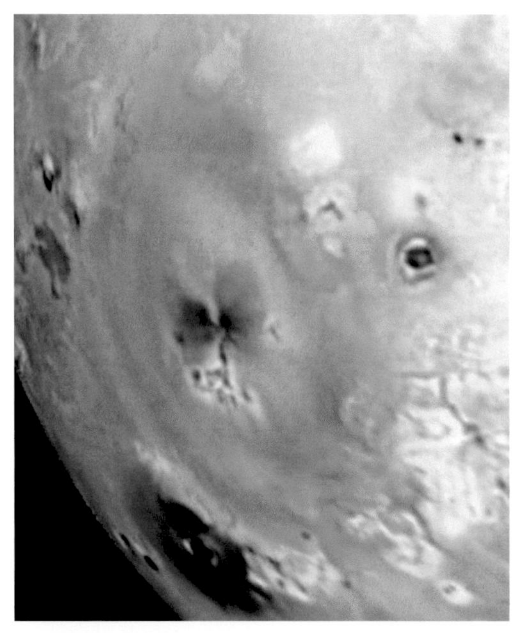

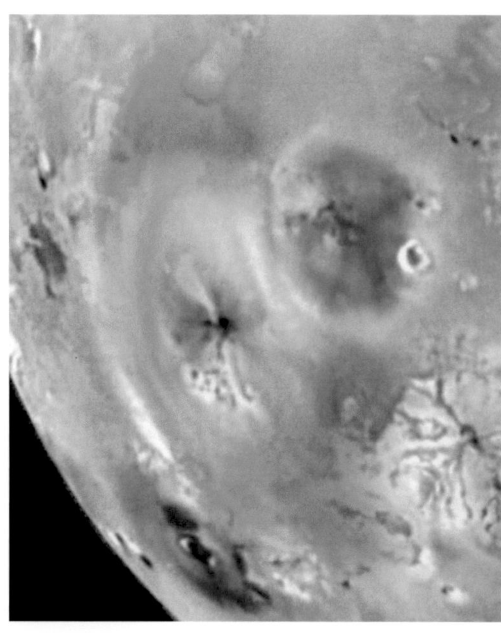

◀ **Ringing the changes:** Major changes on the surface of Io over a five-month period, as monitored by the Galileo space probe. The crimson-coloured ring consists of sulphur compounds, erupted from a volcanic vent at its centre named Pele (after a Hawaiian volcano goddess). The left-hand frame shows the ring's appearance in April 1997. By September (right), a dark spot 400 km (250 miles) in diameter – roughly the size of Arizona – had appeared around a neighbouring vent, Pillan, to its northeast, partly obscuring the Pele ring (Pillan is a South American god of volcanoes). Such volcanic resurfacing happens all the time on Io.

◀ **Hot stuff:** A volcanic eruption in a region on Io called Tvashtar Catena captured by the Galileo space probe. The two small white dots at the left are 'hot spots' where molten rock reaches the surface. The orange and yellow stream is a cooling lava flow more than 60 km (37 miles) long. Other depressions and dark patches are sites of previous eruptions.

▼ **Europa:** The smallest of Jupiter's Galilean satellites is encased in a bright, icy shell, possibly underlain by an ocean of liquid water. Long fractures in the ice are stained reddish-brown by minerals seeping out from the interior. The bright-rayed impact crater at lower right is named Pwll, after the Celtic god of the underworld; it is one of the few impact craters on Europa.

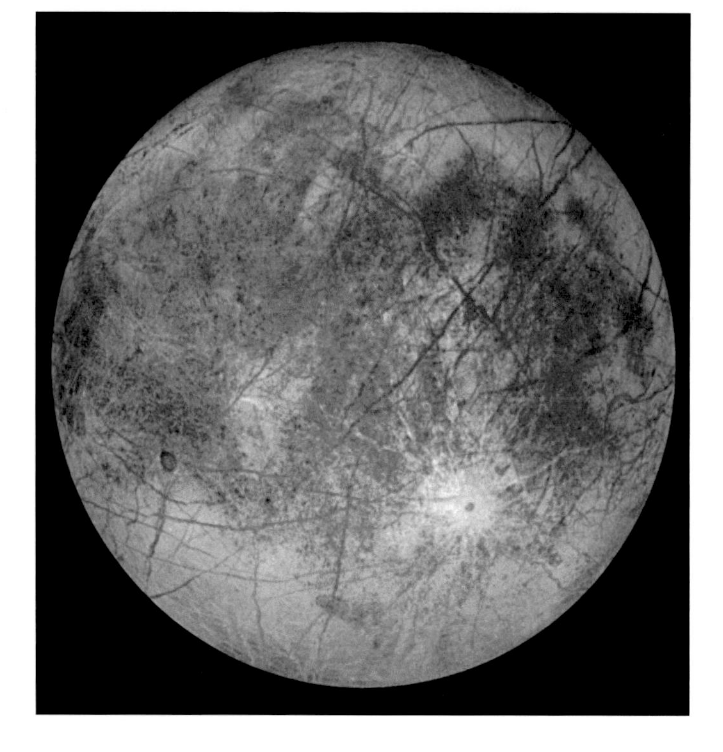

▼ **Going with the floe:** Jumbled ice floes in the Conamara region of Europa resemble the break-up of pack ice on Earth, suggesting the presence of subsurface water either now or at some time in the past. The white-and-blue area at left is covered with crushed ice ejected from the impact crater Pwll some 1000 km (600 miles) to the south. Colours have been exaggerated for clarity. This Galileo image covers an area 53 km (33 miles) wide.

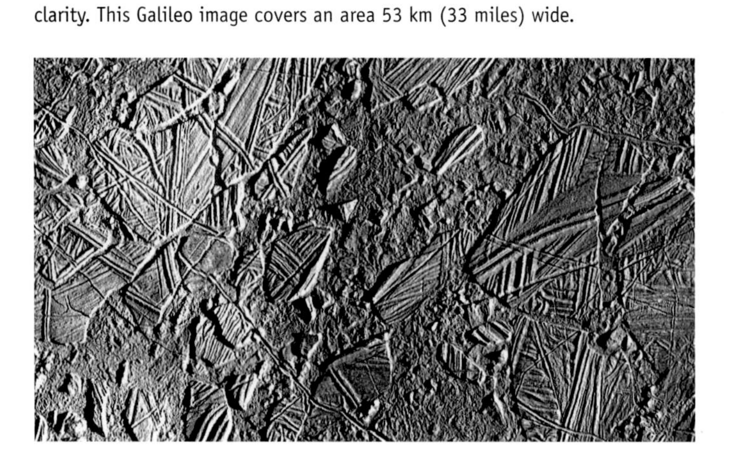

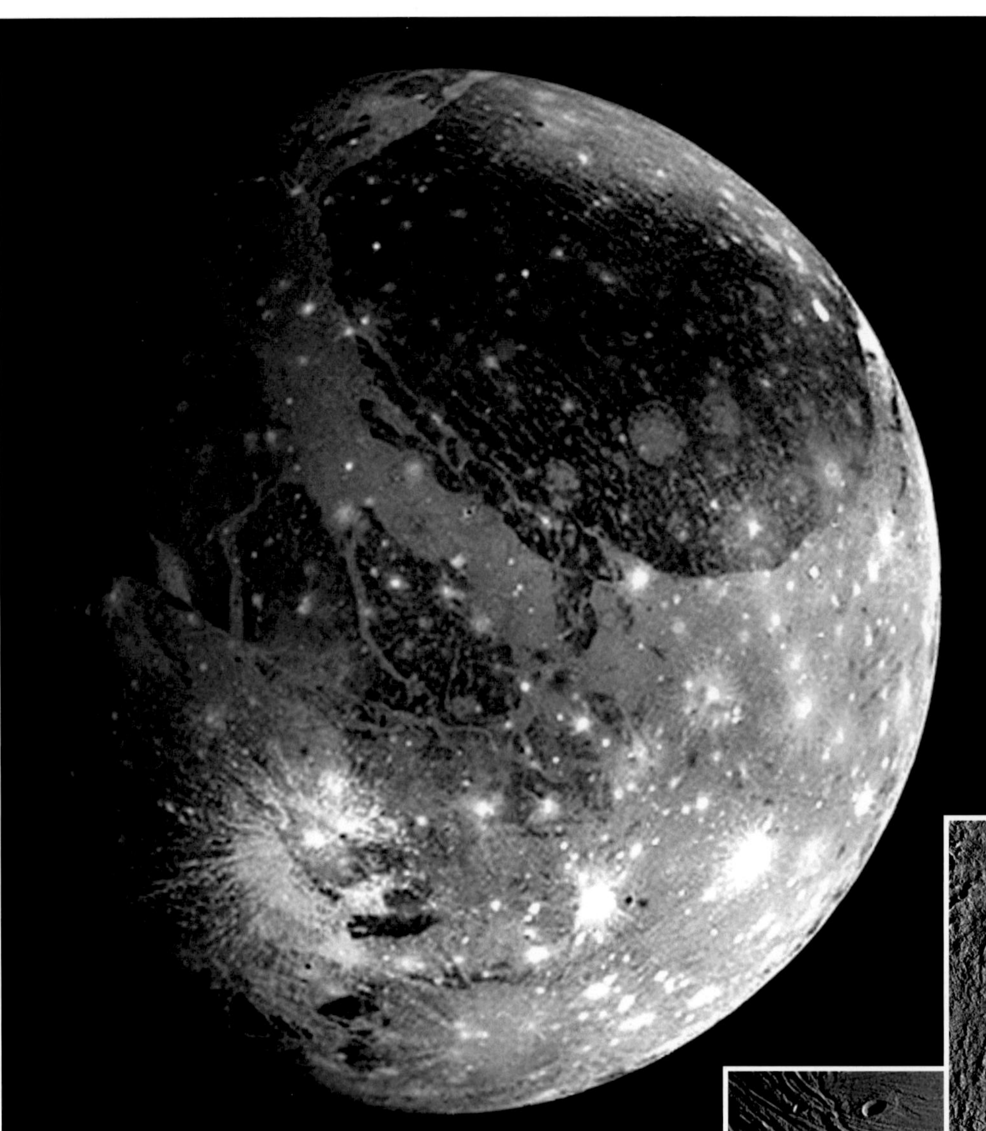

◀ **Ganymede:** Jupiter's largest moon is a ball of rock and ice with a variegated brownish-grey surface. The rounded dark feature in the northern hemisphere is Galileo Regio, over 3000 km (2000 miles) across, an area of heavily cratered ancient crust. Lighter-coloured areas are younger, evidence of extensive surface faulting and ice flows during its history. Below centre are bright splashes of ice exposed by recent impacts. This portrait was taken by the Galileo space probe.

▼ **Wipeout:** The eroded outline of an old impact crater on Ganymede, smoothed out by the gradual flow of underlying ice. Such features are termed palimpsests, after ancient manuscripts that have been partly erased and written over. The younger crater at centre right is 20 km (12 miles) wide.

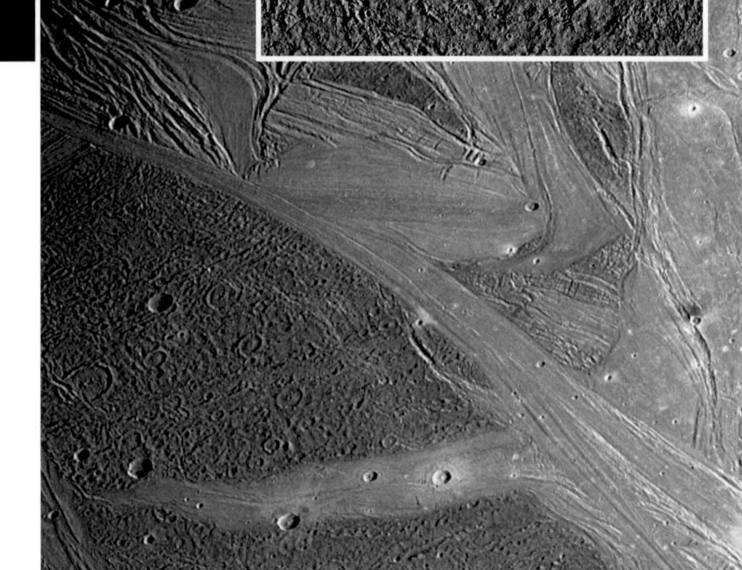

▶ **Two tone:** Unlike on our own Moon, darker areas of Ganymede's surface are not younger lava flows but are ancient and heavily cratered. In places the crust has been stretched and resurfaced by water or slushy ice that has risen from the interior and refrozen, creating smoother, lighter-coloured terrain.

▼ **Rough and smooth:** Three contrasting types of terrain side-by-side on Ganymede. Running like a brushstroke down the centre is a young, smooth strip, caused by faulting and resurfacing with ice. To the right is old terrain, more densely cratered. To the left is terrain of intermediate age, extensively rutted. The total width of this Galileo picture is about 70 km (43 miles).

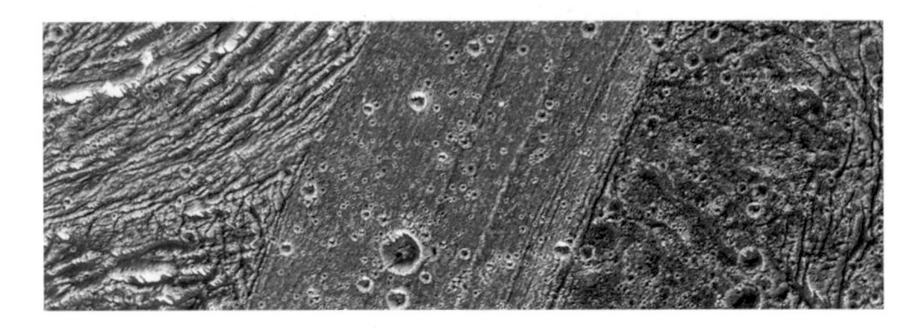

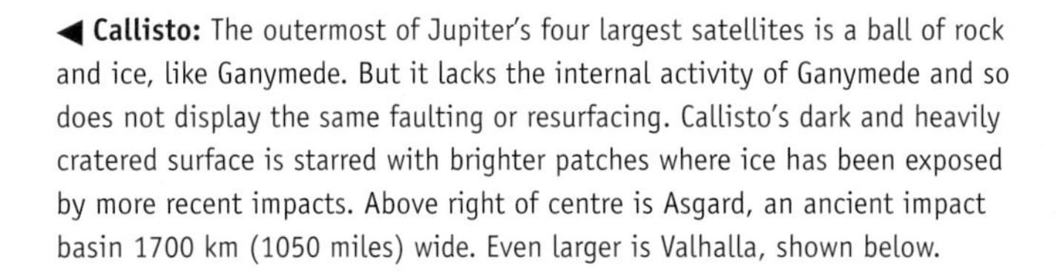

◀ **Callisto:** The outermost of Jupiter's four largest satellites is a ball of rock and ice, like Ganymede. But it lacks the internal activity of Ganymede and so does not display the same faulting or resurfacing. Callisto's dark and heavily cratered surface is starred with brighter patches where ice has been exposed by more recent impacts. Above right of centre is Asgard, an ancient impact basin 1700 km (1050 miles) wide. Even larger is Valhalla, shown below.

▼ **Bull's-eye:** Valhalla, the scar of an ancient asteroid or comet impact, is the largest feature on Callisto. It consists of a bright inner region about 600 km (360 miles) in diameter surrounded by concentric rings 3000 to 4000 km (1800–2500 miles) across, giving it the appearance of a bull's-eye. This oblique view was taken in 1979 by Voyager 1.

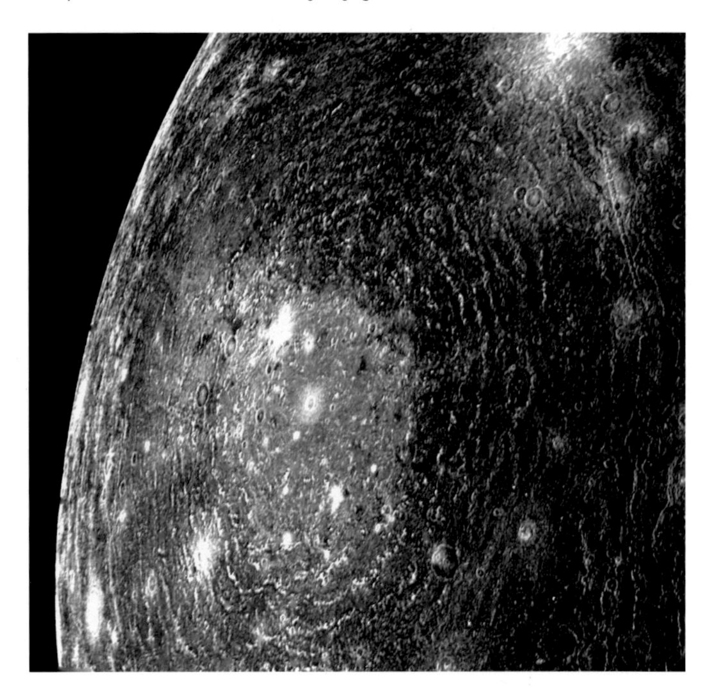

◀ **Uplifting:** Har, an impact crater on Callisto 50 km (30 miles) wide, sports a large central mound due either to rebound of the icy crust at the time of impact or a more gradual upwelling of ice later on. The younger crater on its rim is 20 km (12 miles) across. Diagonal grooves are caused by ejecta from a crater off to the right of picture.

THE DIAPHANOUS DUST RINGS OF JUPITER

Jupiter is surrounded by a series of faint, dusty rings that are invisible from Earth but whose existence has been revealed by photographs from space probes. The main ring, shown below, is about 7000 km (4300 miles) wide. Within it orbit two small moons, Adrastea and Metis, whose surfaces are thought to be the source of the ring's dust, which is kicked off the small moons when they are struck by micrometeorites. At its inner edge the main ring merges into the halo, a much fainter and broader doughnut of material that extends to within about 30,000 km (20,000 miles) of the planet's cloud tops. Just outside the main ring is the broad and exceedingly faint gossamer ring, composed of dust from two other small moons, Amalthea and Thebe. Unlike Saturn's rings, there is no sign of ice in the rings of Jupiter, which is why they are so much darker.

◀ **Dusty ring:** Jupiter's main ring seen almost edge-on by the Galileo space probe. The much fainter gossamer ring lies off to the right, while to the left, closer to the planet, the dusty ring particles spread out above and below the main ring to create a broad but tenuous halo.

Saturn, the ringed world

The most distant planet known to ancient astronomers has turned out to be the most beautiful. Saturn is a planetary icon: a softly textured globe of gas encircled by a flat ring of icy debris. Rings were long thought to be unique to Saturn. Now, Jupiter, Uranus and Neptune are known to possess them too, but those of Saturn remain unsurpassed in brilliance and beauty. If mighty Jupiter is the king of the planets, elegant Saturn is his queen.

Jupiter and Saturn are thought to have been the first planets to form in the Solar System, when gas was still plentiful in the cloud around the young Sun. As a result, these two giant worlds are much the same in composition and internal structure. Through a telescope the most obvious difference between them, rings excluded, is that the markings in Saturn's clouds are more subdued, with only a hint of the bands and spots that decorate the face of Jupiter. This difference is due to temperature: Saturn, being farther from the Sun than Jupiter, has a colder atmosphere. Consequently, its clouds condense lower down, where they are masked from external view by overlying haze.

Saturn is unique in being the only planet with an average density less than that of water. This surprising circumstance arises because it has less than one-third the mass of Jupiter so the resulting pressures in its interior are not as extreme. In theory, Saturn would float if it could be placed upon a sufficiently large ocean. Yet, like a ship, there are parts of Saturn that are far denser than the average: deep within, its hydrogen is compressed into liquid and at the core it is solid and rocky, just like Jupiter. Spinning every 10¼ hours, low-density Saturn bulges more than any other planet, its diameter at the equator being 10% greater than from pole to pole.

Of course, Saturn's defining feature is its ostentatious system of equatorial rings, a swarm of orbiting debris that ranges in size from dust specks to blocks as large as a house. As seen from Earth, the rings divide into three main parts. The outermost, the A ring, is separated from the broadest and brightest central part, the B ring, by a gap as wide as the Atlantic Ocean named Cassini's Division after the 17th-century astronomer who discovered it. Extending inwards from the B ring is the faintest section, the transparent C ring, also known as the crepe ring.

Five Earths could be laid in line from the inner edge of the C ring to the outer edge of the A ring. Yet for all their radial extent, the rings are remarkably thin, no more than a few hundred metres from top to bottom – a million-to-one ratio of width to thickness that makes them proportionally far thinner than a sheet of tissue paper.

Earth and Saturn compared

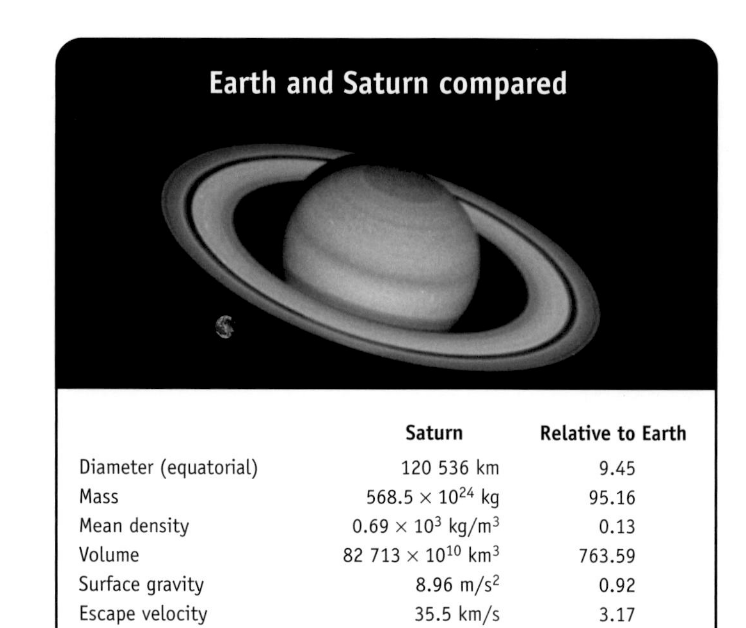

	Saturn	Relative to Earth
Diameter (equatorial)	120 536 km	9.45
Mass	568.5×10^{24} kg	95.16
Mean density	0.69×10^{3} kg/m^3	0.13
Volume	$82\ 713 \times 10^{10}$ km^3	763.59
Surface gravity	8.96 m/s^2	0.92
Escape velocity	35.5 km/s	3.17
Axial rotation period (sidereal)	10.23 h	0.43
Axial inclination	26.73°	1.14
Distance from Sun (average)	1433.53×10^{6} km	9.58
Eccentricity of orbit	0.056	3.38
Orbital period (sidereal)	10 759.2 days	29.46

THE WHITE STORMS OF SATURN

Saturn has no equivalent of Jupiter's Great Red Spot, but white storm clouds well up every 30 years or so, during summer in the planet's northern hemisphere (Saturn orbits the Sun once in about 30 Earth years). Large white spots were seen in 1933, 1960 and 1990, the first and last of these spreading out to fill the planet's equatorial zone before dissipating. A smaller outbreak occurred in 1994 and was photographed by the Hubble Space Telescope (left). Like summer thunderstorms on Earth, Saturn's white spots are caused by rising currents of warmer air. Unlike on Earth, though, the white Saturnian storm clouds are composed of ammonia ice crystals. As seen here, Saturn's winds are eating in to the western edge of the cloud, giving it an arrowhead shape.

SATURN SEEN BY VOYAGER

Serene Saturn displays subtle banding in its ochre-coloured clouds, contrasting with the frenzied meteorology of storm-tossed Jupiter. This portrait was taken by Voyager 2 as it approached the ringed world in 1981. Behind the planet, Saturn's shadow falls across the encircling rings. The darker lane bisecting the rings is Cassini's Division, a clearing through which Saturn's globe can be glimpsed.

This long-range view begins to resolve fine structure in the rings which becomes spectacularly sharper in Voyager close-ups (see next page). One still-puzzling feature is the dark, smudgy streaks which overlie the bright central B ring and orbit around the planet with it (inset, right). Termed spokes, they are presumed to consist of dust particles levitated above the rings by static electricity.

Seemingly almost in line below the globe of Saturn in the main picture are the moons Tethys, Dione and Rhea; the shadow of Tethys is the black dot on the planet's clouds. A smaller moon, Mimas, can just be detected in front of Saturn above left of Tethys, with its shadow falling on the clouds just beneath the rings. The Voyager views remain the best we have of Saturn, but even better ones are expected when the Cassini space probe arrives there in 2004.

HOMING IN ON SATURN'S RINGS AND MOONS

From Earth, the rings of Saturn seem to constitute a solid platter, but under the close-in scrutiny of NASA's two Voyager space probes they unravelled into thousands of individual strands, each consisting of particles moving like a queue of traffic in a strictly defined lane – even the supposedly empty Cassini Division was found to contain a sprinkling of ringlets. External to the A ring, a narrow outer ring called the F ring was seen to be shepherded by a small moon on either side, evidence that gravitational forces from Saturn's moons play a leading role in shaping the rings, but a complete explanation for their stranded structure remains elusive.

Neither is the rings' origin fully understood. It was long thought that they consisted either of material that never formed into a moon, or were the remains of one or more moons that strayed too close to the planet and were ruptured by its tidal forces. Both mechanisms could have played a part, but it also seems likely that the rings are replenished from time to time by fresh ice from impacting comets.

As well as its rings, Saturn is attended by an extensive family of moons. Prime among these is Titan, the second-largest moon in the Solar System and the only one with a substantial atmosphere. No other moons of Saturn rival Jupiter's Galilean satellites in size.

◀ **Ringlets:** Seen in close-up, Saturn's rings break up into countless narrow lanes in this Voyager 2 photograph. The colours have been exaggerated to bring out variations in chemical composition and particle size.

▼ **Looking up:** The Voyager probes were able to view Saturn from directions not possible from Earth. Here, Voyager 2 is looking up at the underside of the rings, illuminated from above by the Sun. When seen from the unlit side like this, relatively opaque areas like the B ring appear dark, while lightly populated zones, such as the C ring and the Cassini Division, are brighter. The F ring is just visible at the outer rim.

▼ **On edge:** As Saturn orbits the Sun its rings are presented at a range of angles to Earth. When exactly edge-on to us, they almost disappear. In this view, taken by the Hubble Space Telescope in August 1995, the edge-on rings appear as a slender thread of light crossing the planet. Their shadow falls on Saturn because the Sun was above the ring plane. Saturn's largest moon, Titan, at left, also casts its shadow on the planet, while Mimas, Tethys, Janus, and Enceladus cluster around the tip of the rings at right.

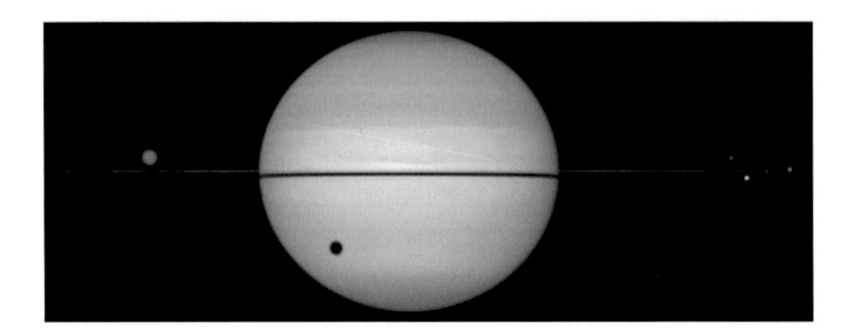

TITAN, SATURN'S SMOGGY MOON

Cloaked in a dense, smoggy atmosphere, Saturn's largest moon, Titan, appeared as merely a fuzzy orange ball to the Voyager 1 and Voyager 2 space probes which flew past it in 1980 and 1981. However, at infrared wavelengths we can see through the smog to the surface, which is how the view at right was obtained with the Hubble Space Telescope in 1994. The bright equatorial area, about the size of Australia, may be a highland of rock and ice. Titan's atmosphere consists mostly of nitrogen (as does that of the Earth), plus a smattering of methane and other hydrocarbons such as ethane; these cause the smog layers and may fall as an oily rain. At the surface of Titan, where the atmospheric pressure is 50% greater than on Earth but the temperature is a frigid -178°C, there may even be lakes of liquid ethane. In 2004 the Cassini probe to Saturn will drop a lander called Huygens onto Titan's surface to experience conditions on this intriguing world.

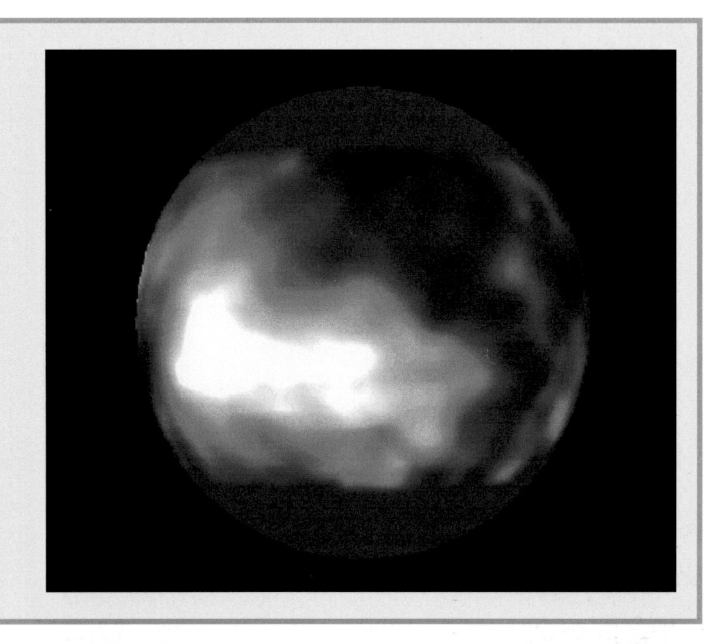

◀ **Mimas:** Resembling the Death Star space station from Star Wars, 392-km (244-mile) Mimas is disfigured by a deep crater more than a quarter its own diameter and larger than Copernicus on our own Moon. The impact that caused it must have come close to shattering icy Mimas.

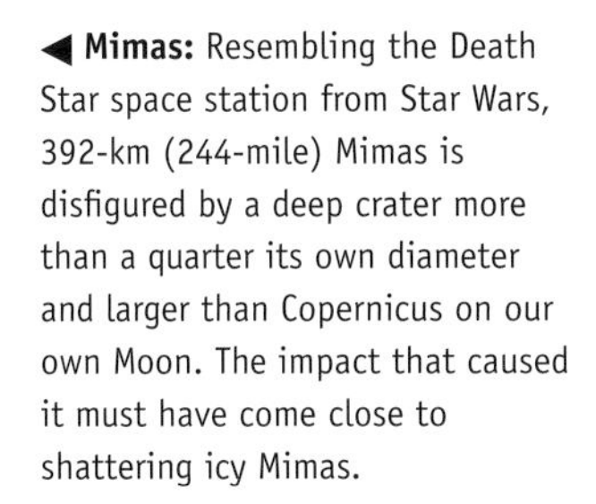

▶ **Tethys:** Another Saturnian moon bearing a major scar, Tethys has a diameter of just over 1000 km (600 miles). The large ring, called Odysseus, 400 km (250 miles) across, probably formed early in the history of Tethys and has since been smoothed out by the flow of ice.

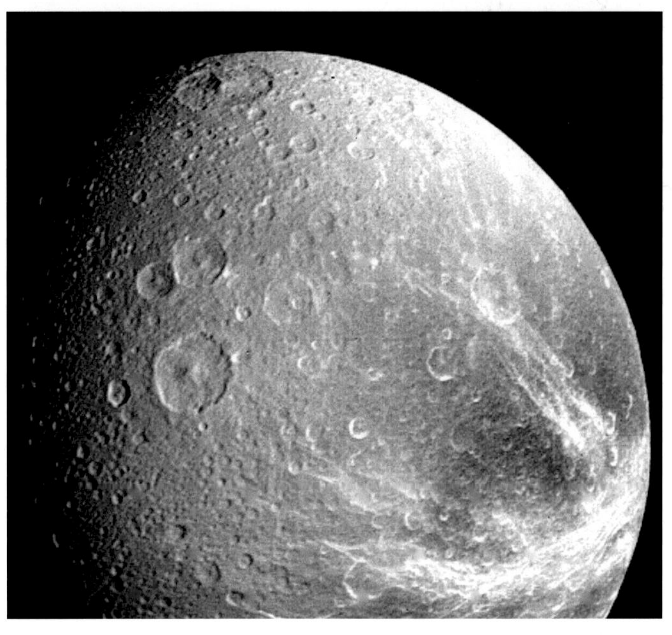

▲ **Dione:** A variety of terrains can be seen on Saturn's fourth-largest satellite, 1120 km (696 miles) in diameter. Various large, shallow impact craters are thrown into relief near the terminator (sunrise–sunset line) in this portrait of Dione taken by Voyager 1 in 1980, while bright, wispy features appear at lower right; these are probably old fault systems.

◀ **Iapetus:** One of Saturn's outer moons, Iapetus is bright on one side and ten times darker on the other, as shown in this Voyager 2 photograph. The dark material may be dust that was knocked off the surface of a more distant moon, Phoebe, and then swept up by Iapetus, coating its leading face.

Uranus, the tilted world

William Herschel, an expatriate German who taught music by day in the genteel English city of Bath and observed the heavens at night through a home-made telescope, was the first person in history to discover a new planet. On the night of March 13 in 1781 he chanced upon a previously unrecorded world that turned out to be four times larger than the Earth and to lie twice as far from the Sun as Saturn, until then the most distant planet known. It was given the name Uranus.

Formed in the colder and less dense outer reaches of the Solar System, Uranus (and also the more remote Neptune, discovered the following century) contains proportionally more water, methane and ammonia than Jupiter and Saturn but a lot less hydrogen and helium. Like Jupiter and Saturn it is a ball of liquid and gas wrapped in clouds, but in this case the clouds appear greenish, due to the presence of methane in the atmosphere overlying them. Visually, Uranus is the least interesting of all the planets: even in the best photographs from Voyager 2, the only probe to have visited it, scarcely any cloud features can be seen.

Uranus seems to have been sideswiped by another large body while it was forming, for its axis of rotation lies almost in the plane of its orbit. Consequently, Uranus has the most extreme seasons of any planet, because the Sun appears overhead at the poles as well as the equator during each 84-year orbit.

Faint, narrow rings discovered from Earth in 1977 encircle the planet's equator, giving tilted Uranus the appearance of a bull's-eye. As well as the rings, Uranus has a family of over 20 moons, the largest of which, Titania, is nearly 1600 km (1000 miles) across.

Earth and Uranus compared

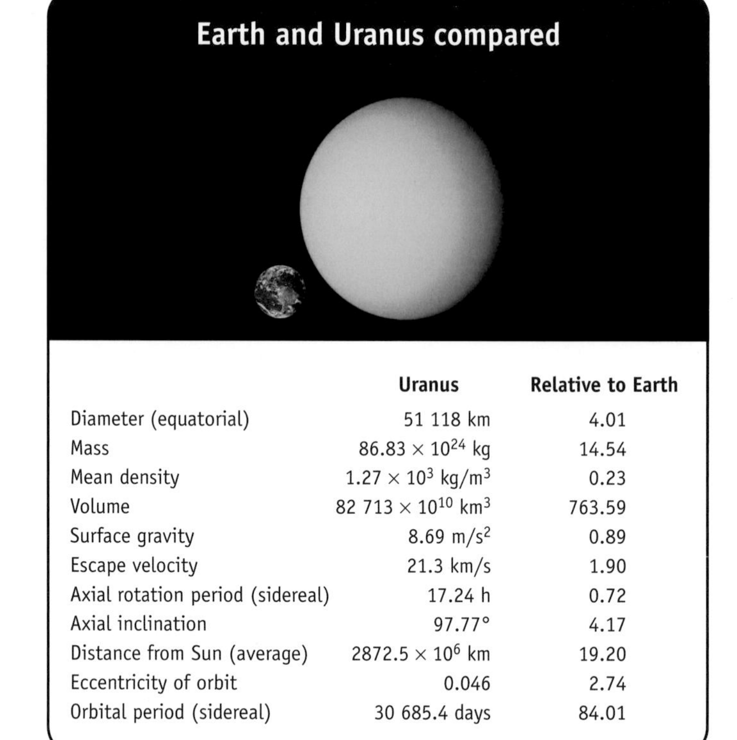

	Uranus	Relative to Earth
Diameter (equatorial)	51 118 km	4.01
Mass	86.83×10^{24} kg	14.54
Mean density	1.27×10^{3} kg/m^{3}	0.23
Volume	$82\ 713 \times 10^{10}$ km^{3}	763.59
Surface gravity	8.69 m/s^{2}	0.89
Escape velocity	21.3 km/s	1.90
Axial rotation period (sidereal)	17.24 h	0.72
Axial inclination	97.77°	4.17
Distance from Sun (average)	2872.5×10^{6} km	19.20
Eccentricity of orbit	0.046	2.74
Orbital period (sidereal)	30 685.4 days	84.01

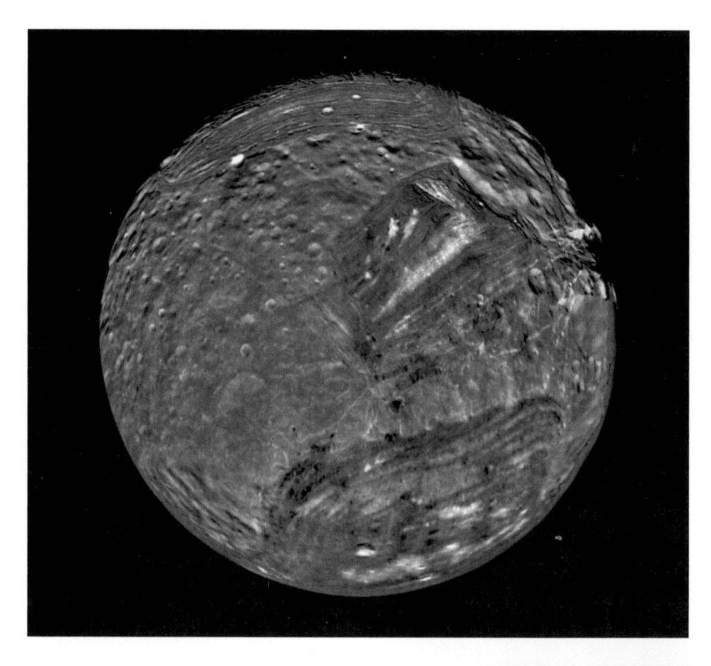

▶ **Miranda:** This astounding moon, although a mere 480 km (300 miles) wide, has one of the most diverse surfaces of any body in the Solar System. At first it was thought to have been broken apart by impacts and reassembled in haphazard fashion, but a subsequent analysis suggests that its features have been shaped by faulting and eruptions of ice.

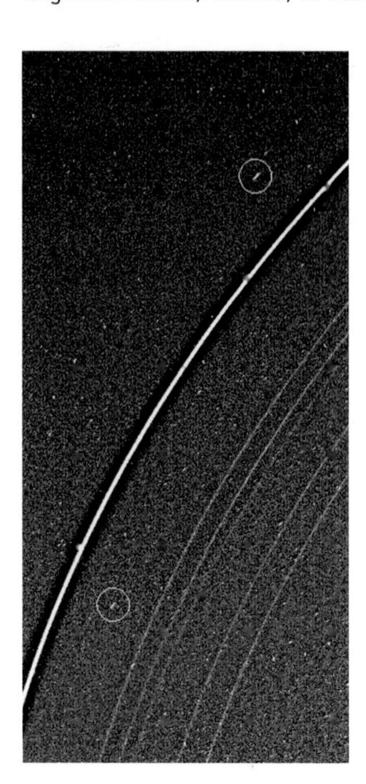

◀ **On guard:** Two small moons (circled), first seen by Voyager 2, orbit like sentries either side of the Epsilon ring of Uranus; they were named Ophelia and Cordelia. Their images are slightly trailed by orbital motion in this Voyager 2 time exposure. Fainter inner rings can also be seen.

▶ **Precipitous:** Sunlight catches the face of a scalloped cliff on Miranda in this close-range view from Voyager 2. The cliff rises up about 5 km (3 miles), higher than the walls of the Grand Canyon on Earth. It can also be seen at upper right on the image above. The impact crater to its left is about 25 km (15 miles) wide.

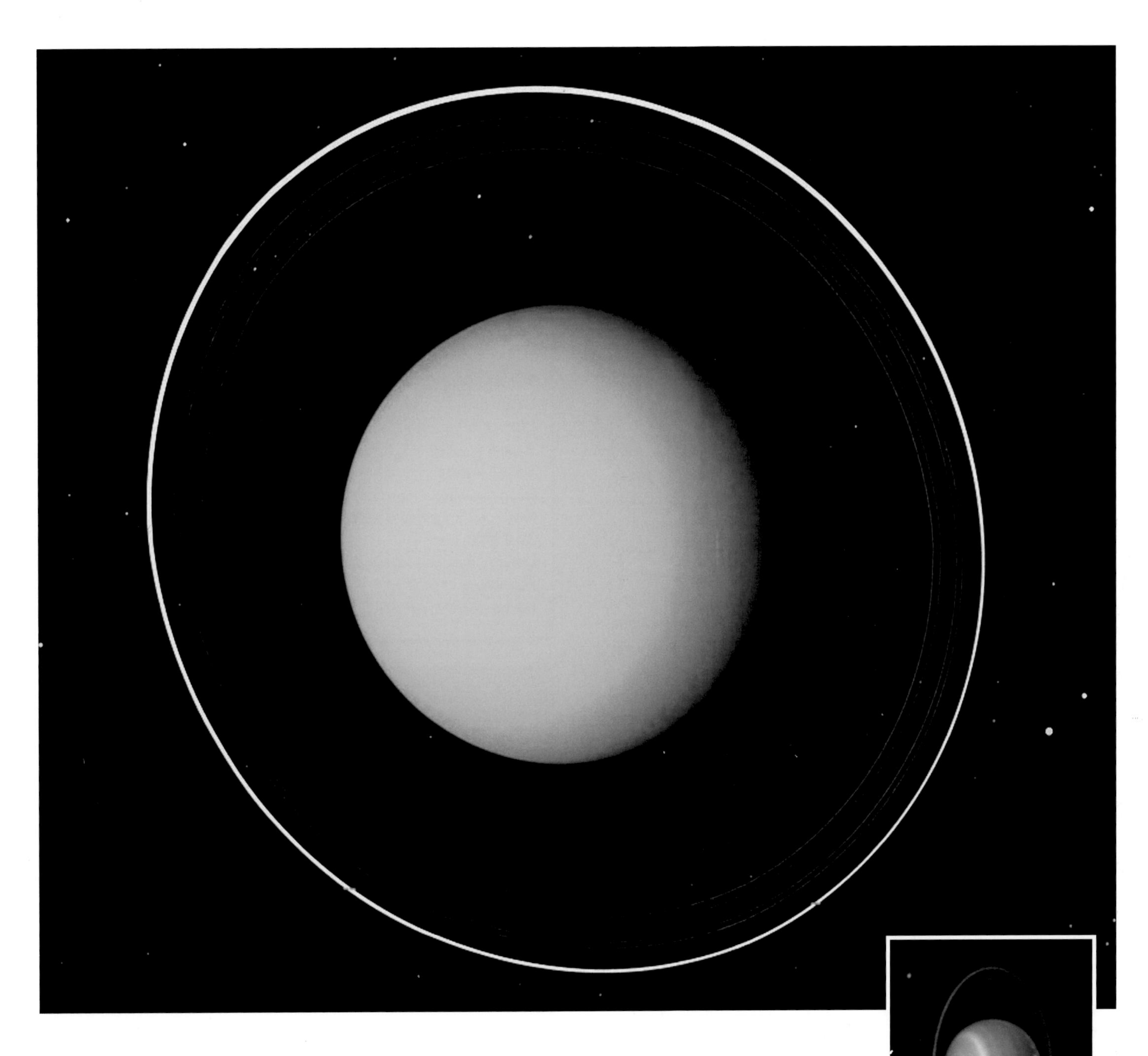

VOYAGER VIEWS A GREEN BULL'S-EYE

Eleven rings circumscribe Uranus, some too faint to see even on this specially processed Voyager 2 portrait. Most prominent of them is the outermost, called the Epsilon ring, 100 km (60 miles) wide. Voyager 2 close-ups showed that the Epsilon ring is shepherded by two tiny moons that orbit either side of it (see facing page). Most of the other rings are only a few kilometres across; it is not yet known whether their narrowness is also due to the regimenting effect of such shepherd moons, too small to have been spotted by Voyager. Whereas the rings of Saturn are icy and bright, those of Uranus are rocky and dark. White dots in this picture are a mixture of background stars and moons of Uranus.

▶ **Spring clouds:** High-altitude clouds in the northern hemisphere of Uranus appear as bright pink spots on this infrared image from the Hubble Space Telescope, presented here in false colour.

The south polar region of Uranus was tipped sunwards when Voyager 2 called in 1986. Cloud features were then almost absent. Twelve years later, as spring sunlight returned to warm the northern hemisphere after its decades-long winter hibernation, the atmosphere burst into life and 20 new clouds were recorded by the Hubble Space Telescope's infrared camera (inset).

Neptune, the outer giant

Observers tracking the motion of Uranus early in the 19th century reported a problem: the newly discovered planet was not keeping to the orbit calculated for it. Either Newton's laws of gravity did not apply at such great distances from the Sun, or else there was another, as yet unseen, planet beyond Uranus, plucking it off course with feeble fingers of gravity. Convinced of the latter, two mathematicians on opposite sides of the English Channel – John Couch Adams in Cambridge, England, and Urbain Le Verrier in Paris, France – set out to calculate where this supposed attractor might lie. Le Verrier finished first and sent his predictions to Berlin Observatory where Johann Galle found the new planet on the night of September 23, 1846.

Neptune and Uranus, almost identical in size, initially appeared to be as alike as two peas in a pod. However, Neptune is bluish because there is more methane gas in the atmosphere above its clouds than on pea-green Uranus. More significantly, Neptune is warmer inside, which stirs up its atmosphere to create cloud features that are largely absent on Uranus. Neptune was seen in close-up in 1989 when Voyager 2 completed its 12-year tour of the four giant planets (Voyager 1 swung upwards out of the plane of the Solar System after passing Saturn, so Uranus and Neptune were never in its sights).

As well as revealing an active atmosphere, Voyager 2 confirmed that Neptune has skimpy rings, similar to those of Uranus. Most remarkable, though, were its views of Neptune's largest moon, Triton, over three-quarters the size of our own Moon and larger than Pluto. Blue and pink expanses of frozen nitrogen and methane were found to cover its blotchy surface, while geysers of gas and dust spouted 8 km (5 miles) high from subsurface pockets of nitrogen. Triton has a very insubstantial atmosphere consisting mostly of nitrogen, which evaporates from the summer side ('summer' here meaning a temperature of –235°C) but freezes again on the winter hemisphere, thereby constantly refreshing the moon's icy surface.

Earth and Neptune compared

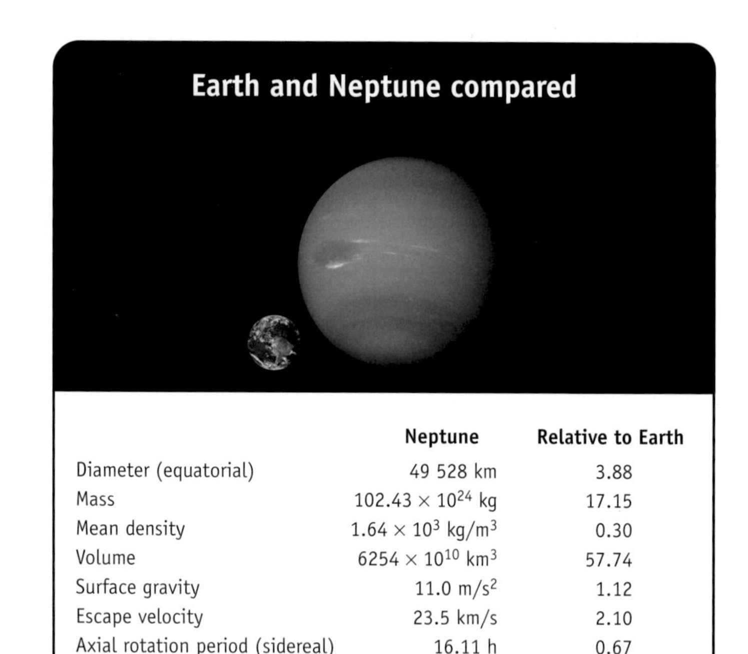

	Neptune	Relative to Earth
Diameter (equatorial)	49 528 km	3.88
Mass	102.43×10^{24} kg	17.15
Mean density	1.64×10^{3} kg/m^3	0.30
Volume	6254×10^{10} km^3	57.74
Surface gravity	11.0 m/s^2	1.12
Escape velocity	23.5 km/s	2.10
Axial rotation period (sidereal)	16.11 h	0.67
Axial inclination	28.32°	1.21
Distance from Sun (average)	4495.1×10^{6} km	30.05
Eccentricity of orbit	0.011	0.68
Orbital period (sidereal)	60 189 days	164.79

◀ **Rings:** The rings of Neptune, backlit by the Sun, photographed by Voyager 2. The three most prominent rings are named for the trio of Neptune discoverers: Adams (the outermost), LeVerrier and Galle (the broad inner ring). Just outside the LeVerrier ring lies a fainter ring divided into narrow and broad components named Arago and Lassell respectively. The overexposed disk of Neptune is partly obscured at centre.

▲ **Triton:** Pinkish ice covers the south polar region of Triton in this Voyager 2 montage. Sooty streaks on the ice are caused by fine, dark dust which erupts from geysers and is then blown downwind for 100 km (60 miles) or so in Triton's thin atmosphere. Around the equator is a bluish-green band thought to consist of nitrogen frost, deposited on an older substratum that is puckered like the skin of a cantaloupe melon.

VOYAGER VIEWS A BLUE LAGOON

When Voyager 2 reached Neptune in 1989, the dominant feature in Neptune's blue clouds was the Great Dark Spot, looking like a large fish basking in a lagoon. A smaller dark spot, shaped like an eye and flecked with bright cirrus at its core, is visible at lower left. White clouds fringing the large dark spot are methane cirrus, as are other bright spots and streaks.

The Great Dark Spot was the Neptunian equivalent of Jupiter's Red Spot, but only about half the size and, as it transpired, nowhere near as long-lived. By the time that the Hubble Space Telescope looked in 1994 it had vanished, as had the smaller dark spot. Instead, Hubble found numerous bright clouds. The two small views

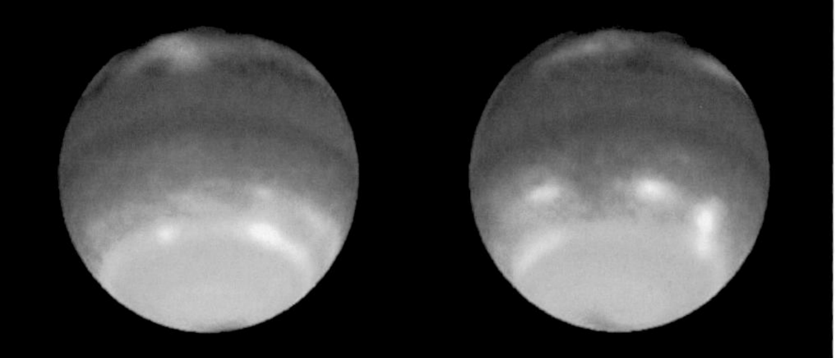

show weather on opposite sides of Neptune in August 1998. Taken through Hubble's infrared camera, they are in false colours; for example, the highest clouds appear pinkish, not white. Neptune has a ferocious equatorial jet stream, with winds of 1400 km/h (900 mile/h), centred on the dark blue belt just south of its equator.

Pluto, the distant dwarf

Pluto is very much the runt of the planetary pack. Indeed, it is so small, only two-thirds the diameter and one-fifth the mass of our Moon, that some astronomers question whether it deserves to be classified as a planet at all. Accentuating its image as a misfit, Pluto renounces the lane discipline of the other planets, following a markedly elliptical orbit that at times moves it within the orbit of Neptune. Pluto was closer to the Sun than Neptune between 1979 and 1999 and will be so again from 2231 to 2245. Fortunately, Neptune and Pluto cannot collide, for their orbits are like interposed hoops that never touch.

Pluto was discovered in 1930 at Lowell Observatory, Arizona, by Clyde Tombaugh, a young assistant hired the previous year to help search for a planet that the observatory's founder, the late Percival Lowell, had believed lay beyond Neptune. Taking pairs of photographs of given regions of sky a few nights apart, and examining them by day in an optical comparator for objects that had moved between exposures, as an orbiting planet would, Tombaugh spent wearying months before he sieved out the nugget he was seeking.

The Lowell Observatory astronomers had expected the new planet to be a giant with an observable disk like Uranus and Neptune, but Pluto remained steadfastly starlike through even the largest telescopes – the first hint of the doubts over its status that were to come. For decades it was assumed to be a dark, rocky body like Mars. Then, in the late 1970s, frozen methane was detected on its surface, indicating that it was actually bright and icy. Next, Pluto was found to have a moon, Charon, which allowed astronomers to calculate its diminutive mass. Pluto's size, composition and orbit all suggest an affinity with the icy asteroids that inhabit the outer Solar System in the so-called Kuiper Belt (see box below). Despite this, the International Astronomical Union, astronomy's governing body, announced in 1999 that there is no intention of changing Pluto's classification as a full-fledged planet.

Earth and Pluto compared

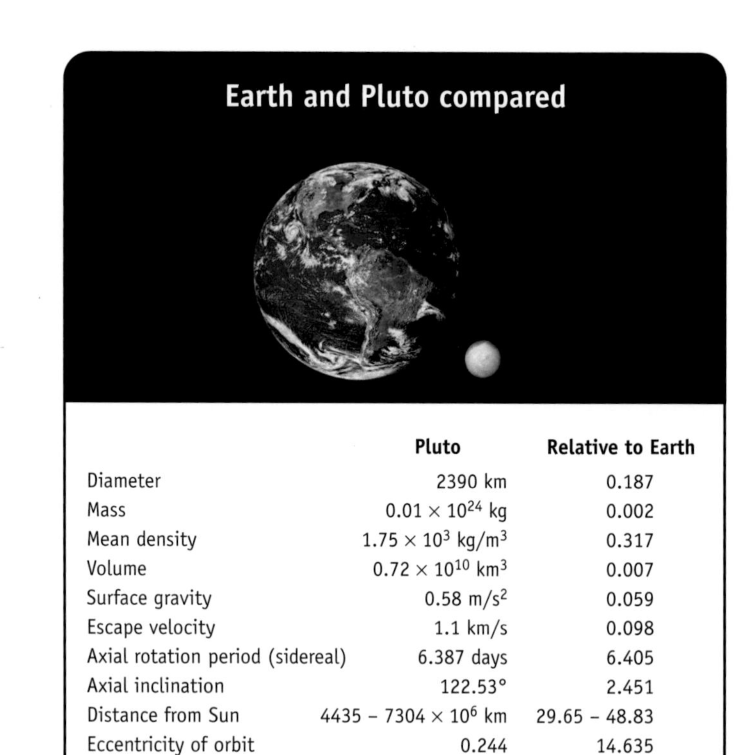

	Pluto	Relative to Earth
Diameter	2390 km	0.187
Mass	0.01×10^{24} kg	0.002
Mean density	1.75×10^{3} kg/m^{3}	0.317
Volume	0.72×10^{10} km^{3}	0.007
Surface gravity	0.58 m/s^{2}	0.059
Escape velocity	1.1 km/s	0.098
Axial rotation period (sidereal)	6.387 days	6.405
Axial inclination	122.53°	2.451
Distance from Sun	4435 – 7304 $\times 10^{6}$ km	29.65 – 48.83
Eccentricity of orbit	0.244	14.635
Orbital period (sidereal)	90 465 days	247.68

PLUTO, PLANET X AND THE KUIPER BELT

After discovering Pluto, Clyde Tombaugh spent a further 12 years searching around the sky for additional planets, without success. In the years that followed, various astronomers raised the prospect that an undiscovered Planet X might lurk at the edge of the Solar System. But such hopes were ended in 1992 by the discovery of the first of a swarm of icy asteroids beyond Neptune. This region is termed the Kuiper Belt – or, alternatively, the Edgeworth–Kuiper Belt, in memory of the astronomers who first proposed its existence. Hundreds of members are now known with diameters of 100 km (60 miles) or more. Numbers increase rapidly at smaller sizes, although they become progressively more difficult to see. Estimates suggest that there may be billions of Kuiper Belt objects in all, their combined mass amounting to several times that of the Earth.

Objects in the Kuiper Belt consist of frozen gas and rock and are the left-overs from the formation of the giant outer planets. They have orbits that extend out to more than 50 times the Earth's distance from the Sun and orbital periods of many centuries. Some mimic Pluto by crossing the orbit of Neptune. The largest known Kuiper Belt object, Quaoar, has an estimated diameter of 1200 km (750 miles), the same as Pluto's moon Charon. Several Kuiper Belt objects have been found to be double, like Pluto and Charon (see picture below). Hence Pluto is now widely regarded as an exceptional member of the Kuiper Belt. In fact, Neptune's largest moon, Triton, which is larger than Pluto, probably once orbited the Sun as a member of the Kuiper Belt before being ensnared by Neptune. The Kuiper Belt is thought to be the source of the majority of short-period comets (see page 60).

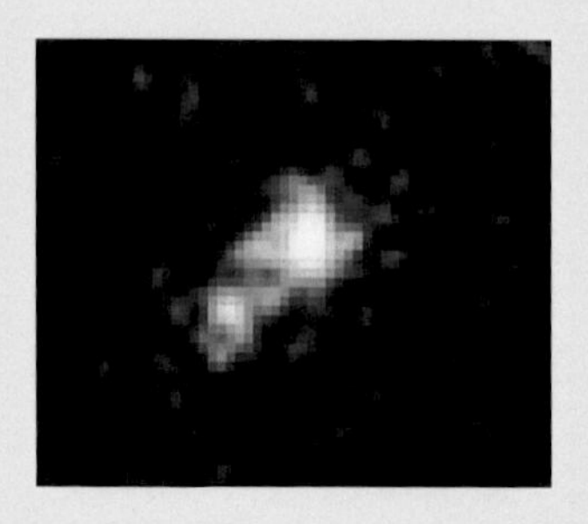

◀ **Twins:** The object known prosaically as 1998 WW31 was the first member of the Kuiper Belt discovered to be double. As the two parts twirl in orbit, the distance between them ranges from 4000 to 40,000 km (2500 to 25,000 miles).

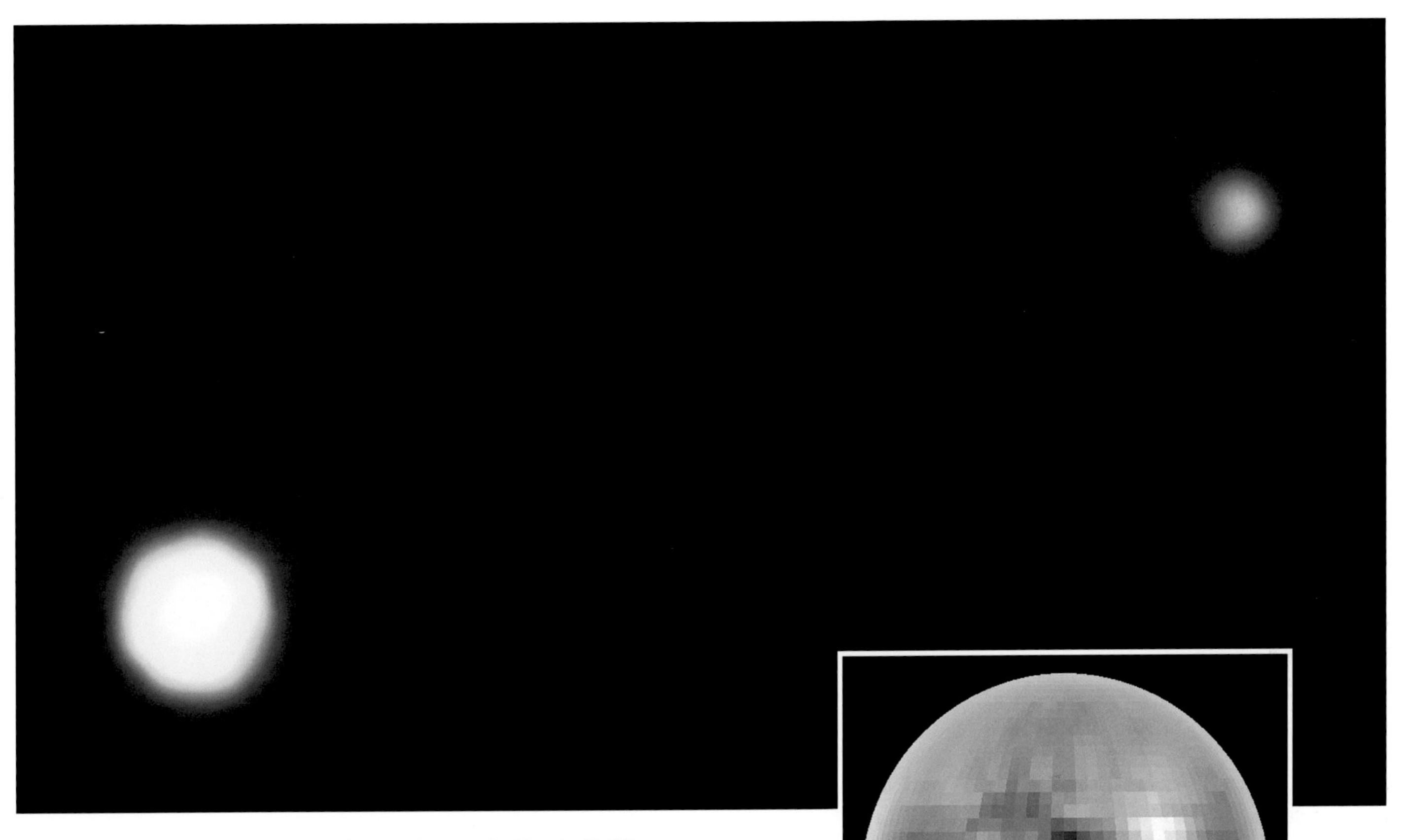

▲ **Double planet:** Pluto and its moon Charon, photographed by the Hubble Space Telescope. Hubble's ability to distinguish Pluto's disk at a distance of 4.4 billion km (2.6 billion miles) is equivalent to seeing the outline of a tennis ball at a distance of 64 km (40 miles). The colour is artificial.

▲ **Pluto in colour:** The true colours of Pluto's surface are shown on this map, which was created from observations of brightness changes while Pluto was being eclipsed by Charon. Pluto's brownish tint is thought to be due to frozen methane which has been discoloured by the action of sunlight.

PLUTO AND CHARON: A DOUBLE PLANET

Pluto's moon Charon, discovered in 1978, is half Pluto's size, a closer match than any other planet-moon pairing (previously the closest similarity was that of the Earth and Moon). Pluto and Charon are therefore regarded as a double planet. Charon orbits Pluto every 6.4 days, the same time that Pluto takes to turn on its axis. As a result, Charon hangs above one point on Pluto like a geostationary satellite over the Earth, permanently visible from one hemisphere of Pluto but never seen from the other. Charon may consist of ice thrown off Pluto in a collision with another body, which - as in the case of Uranus - could also account for Pluto's extreme axial tilt.

From the orientation of Charon's orbit, astronomers deduced that Pluto's axis of rotation is highly tilted with respect to the plane of its orbit. Twice on each orbit of the Sun, Pluto and Charon eclipse each other for a few years at a time as seen from Earth. This last happened from 1985 to 1990, allowing astronomers not only to measure the sizes of the two bodies accurately but also to make rough maps of their surfaces, as shown in the inset. Charon's surface is darker than that of Pluto, which indicates the presence of some material other than ice, such as dust.

For a few decades either side of its closest point to the Sun, which Pluto last reached in 1989, its surface becomes warm enough for some of the ice to turn to gas, creating a tenuous atmosphere predominantly of nitrogen, like that of Neptune's largest moon, Triton. As Pluto moves away from the Sun, the gas freezes again to create a new coating of ice. Pluto will reach its most distant point from the Sun in 2114.

In close-up, Pluto would probably look much like Triton. Pluto is the only planet not to have been reached by space probe. NASA plans to launch a mission called New Horizons in 2006 which would arrive at Pluto and Charon in 2015 or soon after, then fly on to investigate the Kuiper Belt beyond.

Comets, ghostly visitors

From time to time a ghostly apparition steals in front of the stars, remaining for a few weeks or months before fading away as mysteriously as it arose. The Greeks called these celestial spectres 'hairy stars', *aster kometes*, from which comes our word comet. In ancient times such apparitions evoked fear and loathing, and truly bright comets can still inspire awe even though the glare from terrestrial street lighting dilutes their ethereal glow.

Comets may advertise themselves impressively, but in reality even the largest is of little substance. The only solid part is the nucleus, a dirty snowball of ice and dust usually between 1 km and 10 km (0.6 and 6 miles) across. When heated by sunlight, the nucleus releases gas and dust to form a glowing cloud called the coma, from which a diaphanous tail emerges like a stream of cosmic ectoplasm. Dozens of comets are tracked each year, a mixture of known ones returning and new discoveries, but only every few years does one become bright enough to be seen without binoculars or telescope.

Once, the Solar System would have swarmed with comets, for they were the building blocks from which the outer planets were constructed 4.5 billion years ago. The survivors from that process have been preserved, deep-frozen, in two zones: the Kuiper Belt, which extends outwards from the orbit of Neptune, and the larger and more distant Oort Cloud. The Kuiper Belt consists of objects that were too thinly spread to collect into another planet beyond Neptune. Occasional collisions between bodies in the Kuiper Belt propel icy fragments into the inner Solar System.

The Oort Cloud had a different origin. It was created by the gravitational action of the giant planets, notably Jupiter and Saturn, which flung billions of unused ice-balls into a spherical halo that extends out to about two light years from the Sun, halfway to the nearest star. From time to time, stars pass close to the fringes of the Oort Cloud, stirring up its members so that some plunge in towards the Sun on steepling orbits. Unlike comets from the Kuiper Belt, whose orbits lie close to the plane of the planets, those from the Oort Cloud arrive from any direction.

At first, an infalling comet consists simply of a frozen nucleus. As it approaches the Sun and warms up, gas and dust begin to leak away to form an enveloping coma, which can grow to ten times the diameter of the Earth but is less dense than the air we breathe. Gas and dust are pushed away from the coma by the force of sunlight, creating a pair of tails which, in cases such as the majestic comet Hale–Bopp of 1997, can stretch farther than the distance between the Earth and the Sun. Gaseous tails are straight and thin, with occasional twists like the strands of a rope, while the dust particles can fan out to form a tail like that of a magnificent peacock.

After rounding the Sun, the comet recedes back into the darkness of the outer Solar System. How long before it reappears depends where it originated. Comets from the Kuiper Belt return at intervals of less than a couple of centuries. Those from the Oort Cloud take

SEEING COMETARY DUST STORMS

Dust shed by comets spreads out into the Solar System, and may eventually be swept up by the planets. Every clear night, specks of cometary dust the size of sand grains can be seen burning up high in the Earth's atmosphere, producing sudden streaks of light popularly known as shooting stars but which are correctly termed meteors. Inevitably, the densest swarms of meteoric dust lie along a comet's orbit. When the Earth passes close to the orbit of a comet, as it does on several occasions each year, it is engulfed in a dust storm that produces a meteor shower. During such an event, dozens of meteors can be seen spearing across the sky each hour, ten times the rate of random arrivals. Owing to an effect of perspective, the members of a meteor shower appear to diverge from a small area of sky, known as the radiant, and the shower is named after the constellation in which the radiant lies. The year's premier meteor shower is the Perseids, which stream from Perseus each August as the Earth brushes past the orbit of comet Swift–Tuttle. Another comet, Tempel–Tuttle, the parent of the Leonid meteors of November, releases dense clumps of dust each time it passes the Sun on its 33-year orbit. These can give rise to spectacular meteor storms of many thousands an hour. High levels of Leonid activity were seen between 1998 and 2002.

▼ **Leonid storm:** Bright meteors streak like flaming spears from the head of Leo, the lion, during the Leonid display of November 1999 when activity reached 2000 meteors an hour. The bright star at bottom left is Regulus.

▲ **Phantom:** Comet Hale–Bopp, the brightest comet of recent years, hangs like a phantom in the evening sky over Stonehenge in April 1997. Hale–Bopp previously appeared 4200 years ago, about the time that Stonehenge was being built. The gravitational effects of the planets slightly shortened its orbit last time around, so only 2400 years will elapse before its next return.

longer – up to a million years or more, in extreme cases. Some, though, are reined into smaller orbits by the gravitational attraction of the giant planets. This is believed to have happened to the famous Halley's Comet, a visitor from the Oort Cloud which now orbits the Sun every 76 years on average.

Each time a comet nucleus passes the Sun it loses some of its ice and shrinks, becoming gradually less active. Eventually, its glory days behind it, the exhausted comet nucleus orbits the Sun as a dark and dormant body more like an asteroid. Such a fate is befalling comet Encke, which has the shortest known period of any comet, 3.3 years. Encke has made so many approaches to the Sun that it is now severely depleted and never puts on much of a show.

Because they hold clues to the origin of the Solar System, comets are of particular scientific interest. Several probes are planned to rendezvous with, and land on, the nuclei of comets during the coming decade. One NASA mission, Deep Impact, will even blast a crater in the surface of comet Tempel in 2005.

▲ **Black heart:** Fountains of gas and dust spew from the nucleus of Halley's Comet, photographed by the European Space Agency probe Giotto in 1986. The nucleus was found to be 16 km (10 miles) long and to have a very dark, dusty surface. Halley's Comet will return to the inner Solar System in 2061.

COMETARY CRACK-UPS

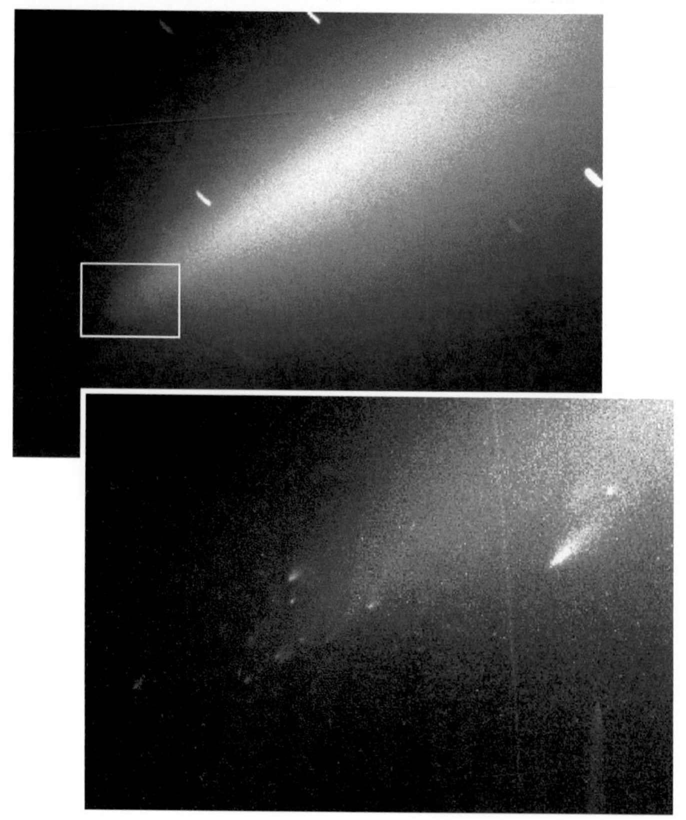

A comet makes a perilously close approach to the Sun, above, as seen by the SOHO (Solar and Heliospheric Observatory) spacecraft. This comet is one of a family called the Sungrazers which follow near-identical orbits that skim the surface of the Sun. Sungrazers are the returning pieces of a super-comet that broke into hundreds of fragments over 2000 years ago. Large ones lip out around the Sun and return to interplanetary space, often rupturing further in the process, as did the brilliant comet Ikeya–Seki of 1965. But others such as the small one above, known only as SOHO-6, are consumed by their fiery passage and disappear. In this image from SOHO's coronagraph instrument, the Sun is masked by a disk; a white circle indicates its true outline. The bright plumes extending from behind the disk are part of the Sun's gaseous corona.

In July 2000 the nucleus of comet LINEAR (named after the Lincoln Near Earth Asteroid Research project which discovered it) vanished after the comet rounded the Sun. Only an elongated tail of dust was visible on Earth-based photographs such as that above right. When the Hubble Space Telescope turned its superior gaze towards the region where the nucleus should have been (the white rectangle), it found a clutch of mini-comets – the remnants of the disintegration (right). Hubble's close-up views support the theory that nuclei are assembled from numerous smaller 'cometesimals'.

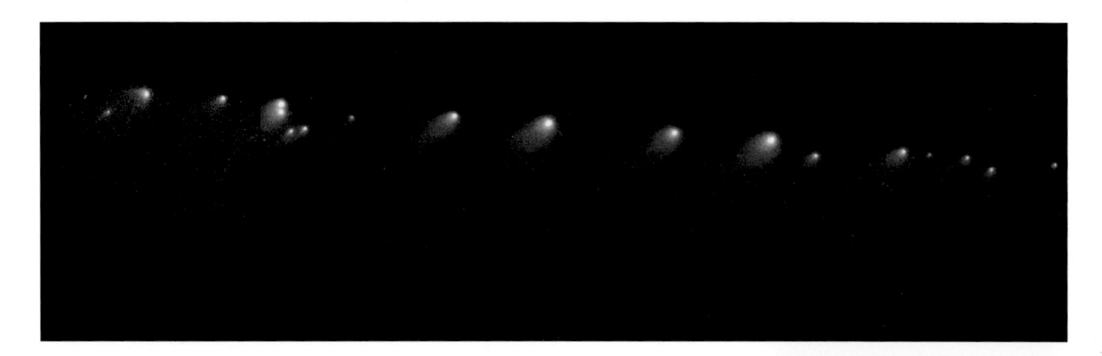

◀ **Gone to pieces:** Comet Shoemaker–Levy 9, broken by Jupiter's gravity into a chain of more than 20 fragments that was likened to a string of pearls, all following the same orbit. The fragments struck Jupiter over a period of a week in July 1994.

THE COMET THAT CRASHED ON JUPITER

The history of the Solar System has been marked by violent collisions, and in 1994 astronomers were able to observe such an episode at first hand as over 20 fragments of comet Shoemaker–Levy 9 plunged into the atmosphere of Jupiter. When discovered in 1993 by the team of Carolyn and Eugene Shoemaker and David Levy the comet's nucleus was already in pieces, forming a row of glowing fragments stretching for three times the distance between the Earth and the Moon. What's more, comet Shoemaker–Levy 9 was found to be in orbit around Jupiter. Calculations showed that Jupiter had captured the comet sometime in the 1920s and that it had broken apart after a particularly close approach in 1992.

One by one, the pieces of the doomed comet peppered Jupiter during a momentous week in July 1994. Tantalizingly, the impacts occurred on the far side of the planet from Earth but, as Jupiter rotated, dark marks in the clouds came into view where each fragment had burned up. Over the ensuing months the markings spread out into a band around the planet that took more than a year to fade. Such events must have been repeated many times during the history of the Solar System. Jupiter's immense size and gravitational influence make it a natural guardian, sweeping up comets and other incoming debris that would otherwise threaten the inner planets, including the Earth.

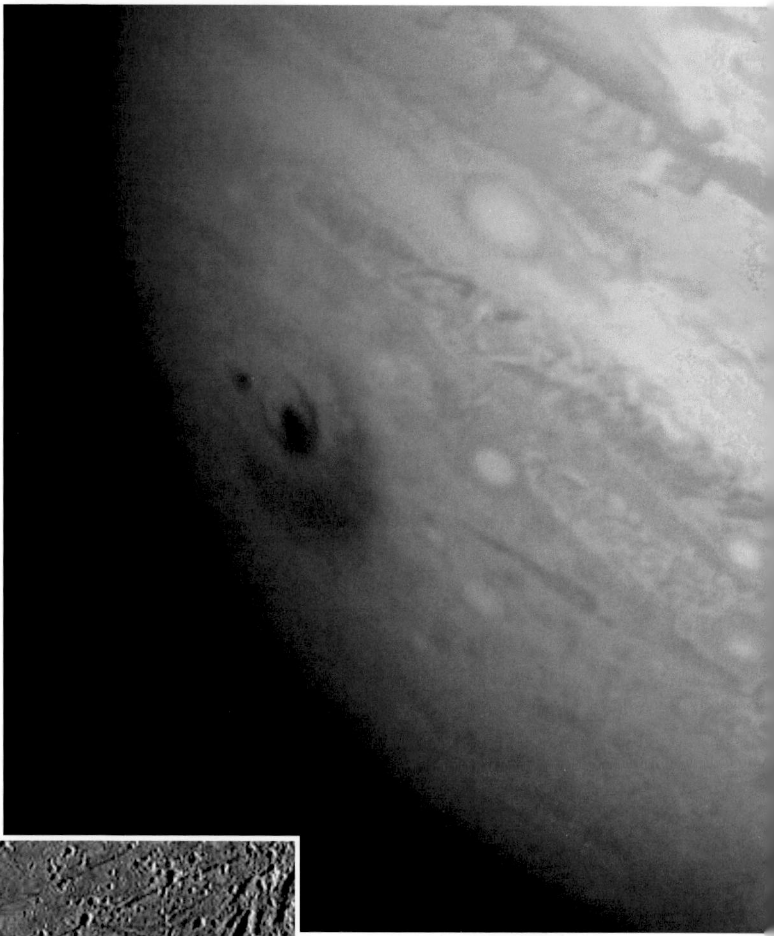

▲ **Black eye:** Looking like a badly bruised eye, this large dark mark in Jupiter's clouds was caused by the impact of fragment G of comet Shoemaker–Levy 9 in July 1994. To its left is a smaller spot caused by fragment D, which landed a day earlier. The central dark spot of the G impact site is 2500 km (1500 miles) in diameter, and is surrounded by a narrow ring 7500 km (4700 miles) across. A fuzzier outer ring is wider than the Earth.

▶ **Impact on Ganymede:** A chain of 13 craters some 150 km (90 miles) long on Jupiter's largest moon, Ganymede, photographed by the Galileo space probe. Similar chains have also been found on Callisto. They are graphic evidence that cometary splits and impacts like that of Shoemaker–Levy 9 have happened in Jupiter's vicinity many times before.

Sun and stars

The midwinter Sun traces out an arc of light from dawn to dusk over the Spanish city of San Sebastian in this 10-hour time exposure. The foreground was captured with a brief exposure before sunrise, while the brilliance of the Sun itself was subdued by a dark filter.

During the day we see only one star, the Sun, whereas at night the sky is strewn with them. Nearest of these nocturnal suns is Alpha Centauri, which lies 4.4 light years away in the southern constellation of Centaurus, the Centaur (a light year is equivalent to 9.5 million million km or 5.9 million million miles). The most distant stars visible to the naked eye are thousands of light years from us, but that is only a paltry fraction of the distances to which telescopes can reach.

Seen from afar, the Sun would appear as a yellow-white star similar to Alpha Centauri, although from our ringside seat on Earth it seems pure white because its brilliance overwhelms our eyes. A star's colour is a guide to its surface temperature. Contrary to our everyday impression that blue is cold and red is hot, the bluest stars are actually the hottest, with temperatures of 30,000°C or more, and the reddest are the coolest, around 3000°C.

In size, stars range from dwarfs one tenth the diameter of the Sun to imposing giants and supergiants dozens or even hundreds of times the Sun's diameter. Our Sun turns out to be a pretty middle-of-the-road star in terms of size, temperature, luminosity and age.

All stars have similar life histories: born from billowing clouds of gas and dust in space, they energize themselves by internal nuclear reactions for between a few million and many billions of years until their internal fuel supply runs out. The main distinguishing factor between stars is their mass. Perhaps surprisingly, the heftiest stars burn out the quickest, exploding after only a few million years in a violent supernova eruption that may leave behind a black hole. More modest stars like the Sun, harbouring their resources for billions of years into sedate old age, fade out gently to a dying ember.

Every star in the sky, like a face in a crowd, presents a snapshot of a life partly run. By assembling their stellar snapshots into chronological order, astronomers can piece together the life story of stars, including that of our own Sun. The following pictures, compressing billions of years of evolution into a few pages, are examples of what they have found.

The Sun, our daily star

At the heart of the Solar System lies a colossal ball of incandescent gas more than 100 times the diameter of the Earth. Whereas planets shine by reflecting light, the Sun – like all stars – is self-luminous, energized to dazzling radiance by nuclear fusion reactions deep within. Hot gas bubbles to its surface, the photosphere, which seethes at a temperature of some 5500°C. Above the photosphere is a more tenuous gas layer, the chromosphere, and beyond that lies the corona, an irregularly shaped halo of gas extending outwards into interplanetary space. The chromosphere and corona are visible only with special instruments or when the blinding photosphere is blotted out at total solar eclipses.

Dusky spots come and go where magnetic fields burst through the photosphere. Because the magnetic fields hamper the outward flow of heat in these regions, sunspots are cooler by up to 1800°C than the rest of the surface and appear dark in contrast with their hotter surroundings. Small spots, about the size of the Earth, last a few days, while large groups persist for weeks or sometimes months. Sunspot numbers rise and fall in a cycle lasting approximately 11 years. Other forms of solar activity also rise and fall in step with sunspots. Examples are the violent but localized eruptions called flares, which fire out high-speed atomic particles that cause radio blackouts on Earth, and coronal mass ejections, immense bubbles of gas blown off from the Sun which give rise to colourful glows called aurorae in the Earth's upper atmosphere.

The source of the Sun's prodigious energy is at its core, where a natural nuclear reactor fuses atoms of hydrogen into helium. Some 600 million tons of hydrogen is converted into helium every second, a process that has been going on since the Sun's birth 4.6 billion years ago. So immense is the Sun that sufficient hydrogen stocks remain to power it for a further few billion years. But eventually the hydrogen fuel will run out and the core will shrink, building up temperatures and pressures sufficiently to initiate the fusion of helium atoms into carbon, a reaction which will increase the Sun's energy output. When this happens the Sun will expand into a red giant, dozens of times larger and hundreds of times brighter than today. As the Sun swells, the Earth's oceans will boil dry and all life on Earth will be extinguished.

Earth and Sun compared

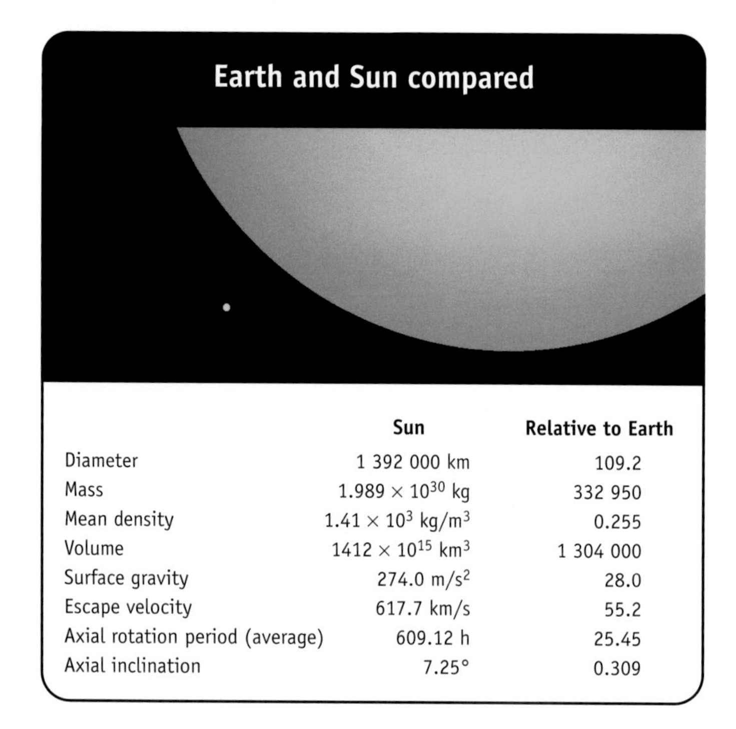

	Sun	Relative to Earth
Diameter	1 392 000 km	109.2
Mass	1.989×10^{30} kg	332 950
Mean density	1.41×10^{3} kg/m^3	0.255
Volume	1412×10^{15} km^3	1 304 000
Surface gravity	274.0 m/s^2	28.0
Escape velocity	617.7 km/s	55.2
Axial rotation period (average)	609.12 h	25.45
Axial inclination	7.25°	0.309

THE SUN AND THE SEASONS

A popular misconception is that the Earth is warmer in summer because we are closer to the Sun than in winter. In fact the opposite is true, at least in the northern hemisphere: the Earth, in its slightly elliptical orbit, comes closest to the Sun in early January, at a distance of 147 million km (91.4 million miles). The intensity of sunlight reaching the Earth is then about 7% greater than it is in early July when we are at our most distant, 5 million km (3.1 million miles) farther away. Because of this difference in distance, the Sun appears slightly larger in January than it does in July. Although too small to be noticeable to the naked eye, the size difference is apparent photographically, as shown at the right. The left side of this composite image shows the size of the Sun in January, while the right half shows it in July. The real cause of the seasons is the 23½° tilt of the Earth's axis. As the Earth orbits the Sun during the year, each hemisphere is tilted in turn towards the Sun. It is summer in the hemisphere tilted towards the Sun but winter in the other.

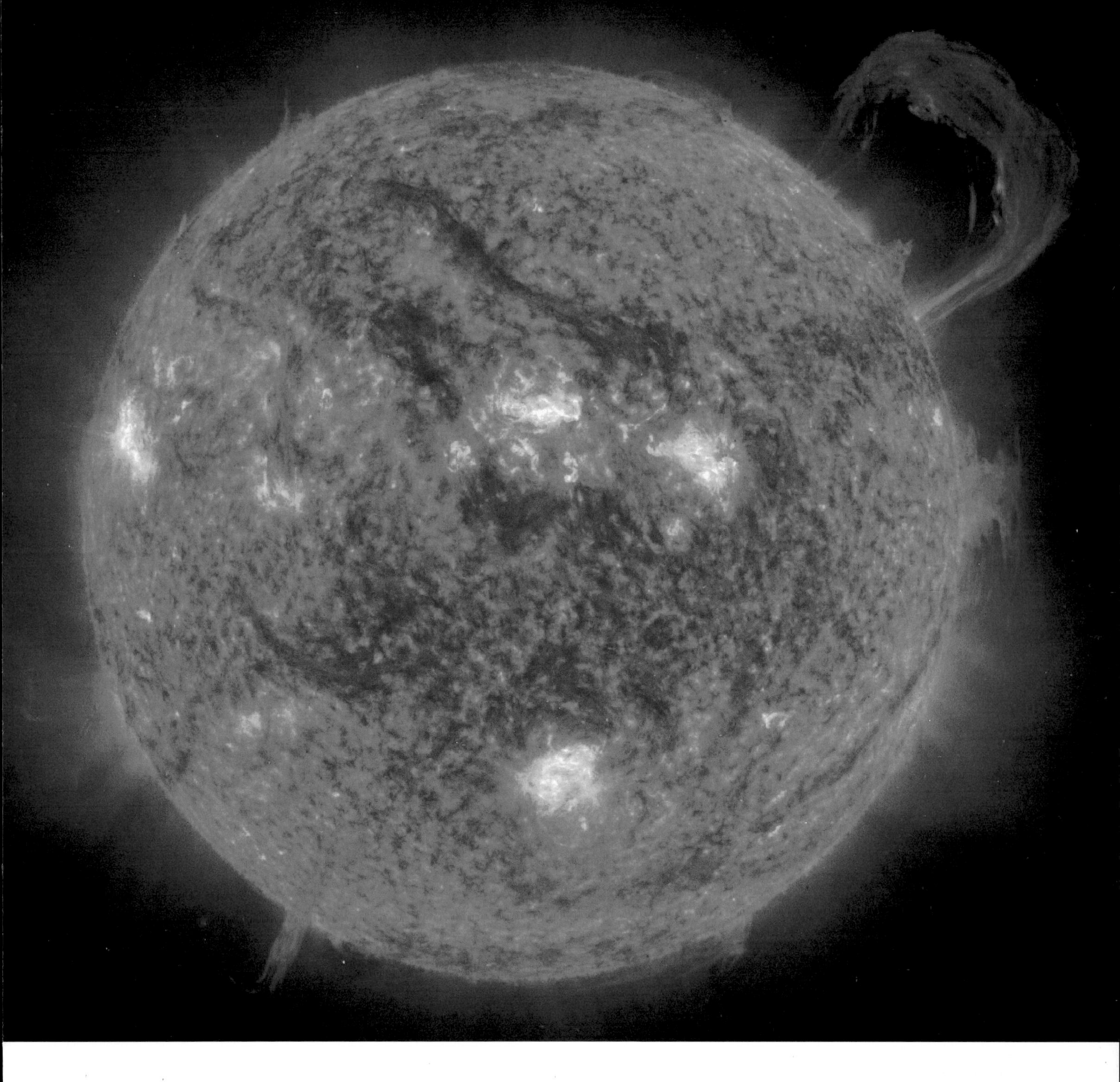

SOHO THE SUNWATCHER

SOHO (the Solar and Heliospheric Observatory) observes the Sun from a point in space 1.5 million km (0.9 million miles) on the sunward side of the Earth, where its view is uninterrupted. The portrait above, taken by SOHO's Extreme ultraviolet Imaging Telescope (EIT), depicts not the surface of the Sun (the photosphere) visible to our eyes but hotter gas a few thousand kilometres above the surface that cannot normally be seen. White areas are hottest while dark red areas are the coolest and map the position of 'holes' in the corona – the source of the fastest currents in the 'wind' of atomic particles (mostly electrons and protons) that blows continually from the Sun. At top right of this image a huge prominence of gas arches some 350,000 km (220,000 miles) into space, sculpted by magnetic lines of force extending out from the Sun. Prominences are denser but cooler than the gas of the surrounding corona, where temperatures reach 2 million degrees. Why the corona should be some 250 times hotter than the Sun's surface is one of the outstanding puzzles of solar physics.

With the help of satellites such as SOHO, astronomers are coming to understand the subtle ways in which various forms of solar activity affect the Earth and its climate. When assessing global warming and its causes, we must not overlook the contribution of the Sun.

SPOTS ON THE SUN

A sunspot resembles a gigantic sunflower: a dark central portion, the umbra, is enclosed by a lighter, petal-like penumbra. Gas flows outwards from the umbra, the coolest part of the spot, into the surrounding penumbra, to be replaced by a downdraught of hotter gas from the corona above. Spots often occur in pairs or larger groups, which in extreme cases can extend for over 200,000 km (120,000 miles), 15 times the diameter of the Earth. Sunspots march across the face of the Sun from day to day as the Sun rotates (right), flowering and then dying back like plants in summer. At times of high solar activity, as in the year 2000, dozens of individual spots and groups speckle the Sun's face at a time, whereas at minimum, as in 1996, spots may be absent for days on end.

Like all forms of solar activity, sunspots are a product of the Sun's magnetic field. At the centre of a sunspot, the magnetic field is thousands of times stronger than that of the Earth. Solar flares can erupt in the tangled web of magnetic fields around complex groups, heating gas to many millions of degrees in just a few minutes and releasing as much energy as a billion megaton bombs. Flares spit out atomic particles that reach the Earth in about a quarter of an hour, having travelled at speeds approaching that of light.

Brighter patches on the Sun's surface called faculae presage the appearance of sunspots, or linger like ghosts after spots have died. Surprisingly, the Sun is slightly brighter at sunspot maximum since faculae are then also more abundant and their brightness more than compensates for the darkness of the sunspots.

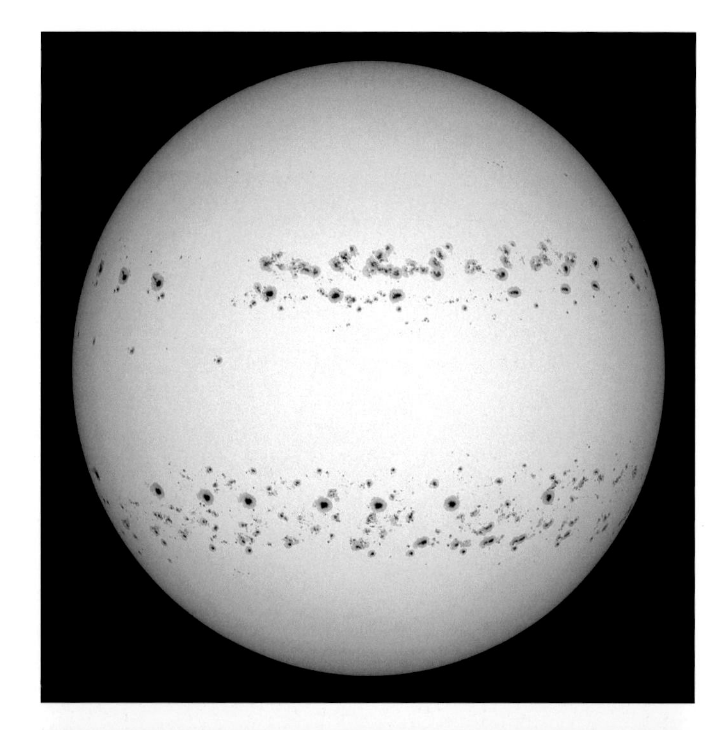

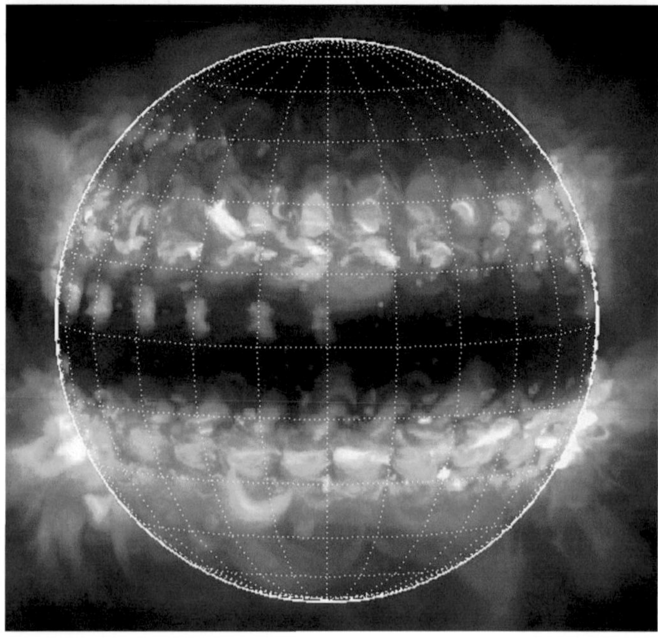

▶ **Spot check:** The Sun's rotation is illustrated in the remarkable composite view at top right, assembled from daily images covering most of August 1999. Spots are carried across the Sun's face by solar rotation, seeming to overlap where they follow each other at similar latitudes. The lower picture shows the Sun for the same month in X-rays, as seen by the Japanese satellite Yohkoh. Loops of hot gas, which follow magnetic lines of force above sunspots, emit strongly at X-ray wavelengths and show up brightly in the Yohkoh images.

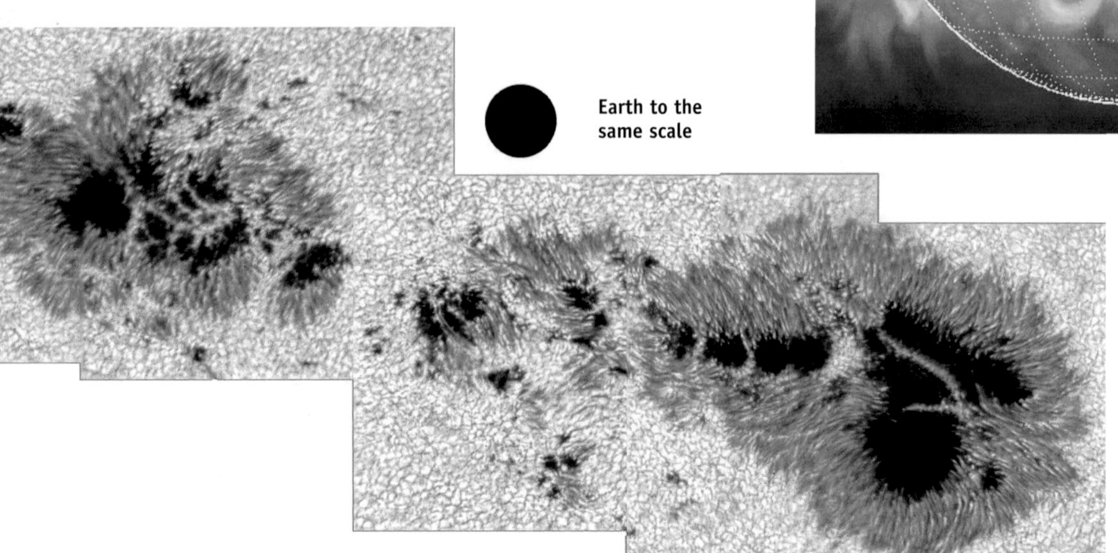

◀ **Sunflower:** This complex sunspot group appeared in September 1998 and stretched for some 200,000 km (120,000 miles). The mottled appearance of the surrounding photosphere, termed granulation, is caused by currents of hot gas rising to the surface. This composite image was taken with the Vacuum Tower Telescope on the Spanish island of Tenerife.

◀ **Ballooning:** A huge bubble of hot gas known as a coronal mass ejection balloons away from the Sun in February 2000, as seen by the Large Angle and Spectrometric Coronagraph (LASCO) instrument aboard the Sun-watching satellite SOHO. LASCO observes the corona out to a distance of 16 times the Sun's diameter, over one-third the way to the orbit of the innermost planet, Mercury. The Sun itself is obscured by a circular disk; its outline is shown by a white circle.

▼ **Glowing:** A colourful aurora photographed from orbit by astronauts aboard the Space Shuttle Discovery in 1991. Spires of fluorescent gas rise up to 500 km (300 miles) above the Earth, the height at which the Shuttle and many other satellites orbit. Discovery's tail is visible at left.

BLOWING BUBBLES IN THE WIND FROM THE SUN

They are the largest eruptions on the Sun but, until the 1970s, the existence of so-called coronal mass ejections (CMEs) was unknown because they cannot be observed from Earth. Spacecraft such as SOHO and its predecessors, equipped with instruments that block light from the Sun's dazzling disk to reveal the fainter corona, have shown that the Sun frequently sloughs off huge bubbles of matter. In a CME, hot gas held down by the magnetic fields that permeate the corona is suddenly released and rises like a balloon. When solar activity is at a minimum, about one such event occurs each week, whereas at maximum activity the number rises to more than one per day. Each CME can carry off a mass equivalent to that of 100,000 fully laden oil tankers.

Expanding away from the Sun at speeds up to 7 million km/h (4.5 million mile/h), CMEs can grow to become many times larger than the Sun itself and create storm fronts in the solar 'wind' of atomic particles. Should a CME be headed our way, the Earth will feel its full force a day or two after the eruption. Guided by the Earth's magnetic field, atomic particles cascade down onto the atmosphere around the magnetic poles, causing atoms and molecules of oxygen and nitrogen to glow colourfully like a fluorescent tube, mostly green and red. Such a display is termed an aurora (also known as the northern and southern lights). Other effects associated with the arrival of a CME include damage to artificial satellites caused by atomic particles, and power failures on Earth from currents induced by the twanging of the Earth's magnetic field. A solar flare sometimes occurs at the same time as a CME. Flares were long thought to be the cause of aurorae and their associated effects, but CMEs are now known to be the true culprits.

▲ **Corona:** A ruffled collar of filmy gases, the corona, fringes the Sun at a total eclipse. Among the most entrancing sights in nature, the corona becomes visible only when the Sun's brilliant surface is blotted out by the Moon. As well as the corona, ruby-coloured prominences of hydrogen can be seen springing into space around the Sun's rim in this portrait of the June 2001 eclipse – the first of the 21st century – taken from Zambia, Africa.

DARKNESS IN DAYTIME

From time to time, the Moon slips in front of the Sun and throws its shadow on part of the Earth. Within this shadow zone, some or all of the Sun is screened from view. From the outskirts of the shadow, the Sun is seen partially eclipsed. But anyone at the very centre of the shadow track experiences one of the rarest and most beautiful spectacles in nature: a total eclipse of the Sun. Eclipse aficionados travel the world in search of the few brief minutes of daytime darkness that totality affords.

Each year, at least two eclipses of the Sun, either partial or total, are visible from somewhere on Earth; the maximum number of solar eclipses in a year is five. A solar eclipse starts almost imperceptibly when the Moon takes a small bite out of the Sun's edge, unnoticed except by careful Sun-watchers. Not until more than half the Sun has been covered are the effects of the eclipse apparent to a casual observer, as the sky begins to darken like a gathering storm. An hour after the eclipse began, for those on the centre line, the Sun has shrunk to a slender crescent and an autumnal chill can be felt in the air. Animals and plants react as though night is falling.

Daylight ebbs away rapidly in the final moments before totality as the Moon rolls in front of the Sun like a boulder across the mouth of a cave. A few chinks of departing sunlight peep through valleys at the Moon's edge, creating the effect known as Baily's Beads; one of the beads may outshine the others, giving the appearance of a diamond ring. But the advancing Moon soon snuffs out even these last glimmers, imprisoning the landscape below in an eerie twilight. In place of the brilliant Sun is a black disk – the silhouette of the Moon – fringed by the pearlescent plumes of the solar corona (above). Bright stars and planets shine out in the darkened sky.

All too soon, though, the spectacle is over. The diamond ring flashes out, to be followed by an ever-expanding crescent of sunlight. The landscape brightens again as though a curtain has been drawn back. Daytime has returned.

Totality: Occulted by the Moon, the Sun hangs like a black hole in the darkened sky above Elazig, Turkey, at the total solar eclipse of August 1999 – the last of the second millennium. Venus shines out 6 cm (2¼ inches) to the left of the Sun.

Stargazing

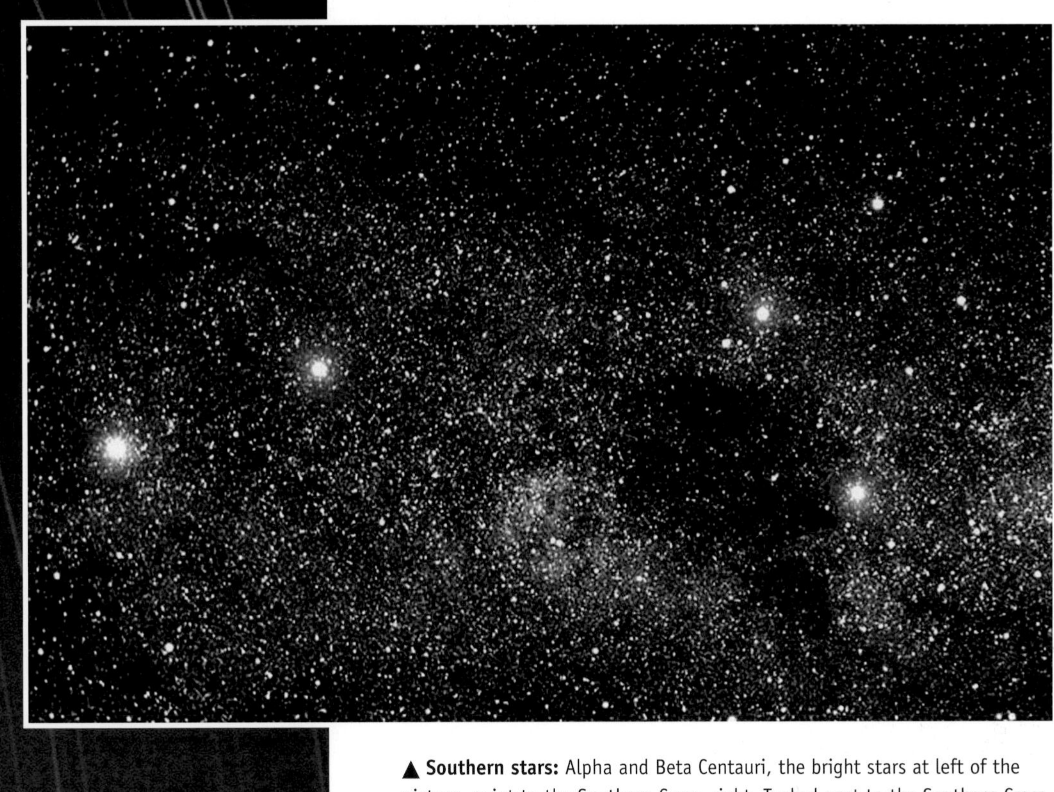

▲ **Southern stars:** Alpha and Beta Centauri, the bright stars at left of the picture, point to the Southern Cross, right. Tucked next to the Southern Cross is the charcoal smudge of the Coalsack Nebula, a cobweb of dust that screens light from the stars of the Milky Way behind. Alpha Centauri, a yellow-white star, is similar to the Sun in temperature and colour, while Beta Centauri is much hotter and noticeably bluer. Such colours are elusive to the eye but binoculars bring them out more clearly. Colour film often exaggerates star colours, as here. Alpha and Beta Centauri are seemingly adjacent on the flattened perspective of a photograph, but in reality Beta is over 100 times more distant. In most cases, stars in a constellation are not related to each other; they appear together only by chance alignment.

◀ **Circling the pole:** On a time exposure, stars trace out arcs around the celestial pole as the Earth spins. A moderately bright star called Polaris lies close to the north celestial pole, to the convenience of navigators in that hemisphere. But the southern hemisphere has no such lode star. This eight-hour exposure of the area around the south celestial pole was taken from the shores of Lake Titicaca, Bolivia, high in the Andes mountains. In the foreground is a replica of Ra, the vessel constructed from papyrus reeds in which the Norwegian anthropologist Thor Heyerdahl crossed the Atlantic Ocean. The two bright arcs above the mast of Ra are Alpha and Beta Centauri, with the four stars of the Southern Cross below and to their right.

Starbirth

Swirling clouds of hydrogen and helium gas called nebulae (Latin for mist) are the birthplaces of stars. Within these interstellar fogs, denser knots of gas begin to shrink under the inward pull of gravity. At the centre of the contracting gas globule, temperatures and pressures climb until they become high enough to spark nuclear fusion reactions that convert hydrogen into helium. Such reactions are similar to those in a hydrogen bomb, the difference being that the weight of a star's overlying layers prevents it from blowing asunder. The onset of nuclear fusion transforms the gas ball into a self-luminous star.

The assembly of gas into stars continues today in nebulae such as that in Orion, faintly visible to the naked eye south of the three stars that stud Orion's belt (see left). This gas cloud, over 20 light years in width, is spawning stars in their hundreds. The brightest new arrivals can be seen through binoculars, lighting up the centre of their foggy nursery. Others, still in the embryo stage, can be detected at infrared wavelengths, since they are not yet hot enough to emit visible light. When we look at present-day star hatcheries such as the Orion Nebula, we are in effect witnessing a replay of how our own Sun was born, 4.6 billion years ago.

Stars come in a broad range of masses, from king size to dwarfs, but there are certain limits. At the top of the range are supergiants with about 120 times the mass of the Sun. This is the maximum that is physically possible – a star with a greater mass would be so hot and luminous that it would literally disintegrate. The smallest stars, the cool and faint red dwarfs, contain a mere 8 per cent the mass of the Sun (equivalent to 80 times that of the planet Jupiter). Gas balls with masses just below this threshold of true stardom never achieve nuclear ignition. Instead, they become so-called brown dwarfs, substars that glow feebly with the heat generated by their shrinkage before gradually fading out.

◀ **Orion Nebula:** About 1500 light years away in the constellation of Orion, this star-forming cloud of gas displays itself like the bloom of an exotic orchid. The nebula's flushed-pink colour comes from hydrogen gas, which blends with green emission from oxygen in the centre to produce a yellowish hue. Within the brightest part of the Orion Nebula, near a wedge of darker gas and dust, is a clutch of new-born stars called the Trapezium, seen in close-up below. This portrait was taken with the Anglo-Australian Telescope.

▼ **New arrivals:** Four bright young stars form a trapezium shape at the centre of the Orion Nebula, seen below in this Hubble Space Telescope close-up taken in visible light. When the same area is viewed with Hubble's near-infrared camera, below right, dozens more bright points spring into view. These are embryo stars and brown dwarfs, either obscured from view at visible wavelengths by surrounding nebulosity or too cool to emit visible light.

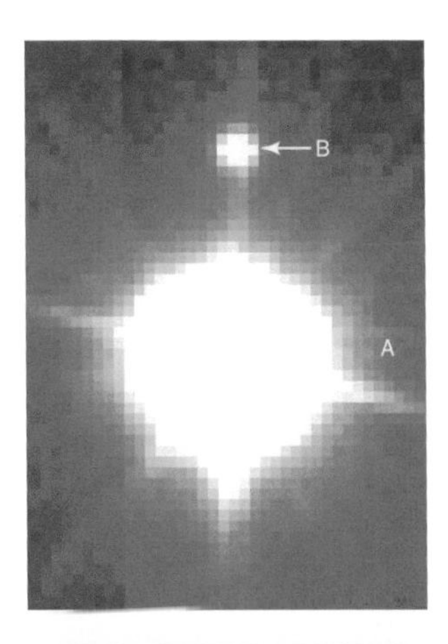

◀ **Substar:** A brown dwarf known as TWA-5 B, arrowed, orbits the much brighter normal star below it, labelled A. TWA-5 B has an estimated mass only 15 to 40 times that of Jupiter, too small to be a proper star but still too large for a planet. The brown dwarf and its companion are part of a field of young stars about 180 light years away in the constellation Hydra. They were imaged by the European Southern Observatory's Very Large Telescope in Chile.

▶ Black knight: Bearing an irresistible resemblance to a knight in a celestial chess game, the Horsehead is the most distinctive nebula in the heavens. It consists of a cold cloud of gas and dust silhouetted against a backdrop of glowing hydrogen. The surface of the Horsehead is being sculpted and eroded by the light from the bright star Sigma Orionis, off the top of the picture. Like the famous gas columns in the Eagle Nebula, seen on the facing page, the Horsehead Nebula swaddles infant stars, one of which can be seen at its top left edge. This false-colour image was taken to mark the 11th anniversary of the Hubble Space Telescope. The Horsehead Nebula, shown in a wider view on pages 2–3, lies in Orion but is not part of the larger Orion Nebula.

◀ Home and away: Star formation in other galaxies as well as our own can now be studied in detail thanks to the power of the Hubble Space Telescope. The nebula shown here, NGC 604, lies in one of the arms of the spiral galaxy M33, around 3 million light years from us in the constellation Triangulum. NGC 604 is immense, around 75 times the diameter of the Orion Nebula. At its heart is a litter of over 200 young stars with masses up to 60 times that of the Sun. Their intense radiation has hollowed out the nebula's centre, exposing them to external view. Like lanterns in a cave, they now light up the surrounding folds of gas and dust, which are here depicted in false colour, coded to emphasize temperature differences – in reality the gas is pink and the hot stars are blue.

EGGS IN THE EAGLE'S NEST

Gnarled columns of cold gas and dust project into the heart of the Eagle Nebula, 7000 light years away in the constellation Serpens, as seen by the Hubble Space Telescope. Ultraviolet light from hot young stars off the top of the picture illuminates and erodes the surfaces of the columns, exposing nodules of denser gas dubbed EGGs (Evaporating Gaseous Globules) which lie within the finger-like protrusions at the end of the columns. Stars are forming inside some of these EGGs and will emerge when the gas around them has evaporated. In this false-colour image, hydrogen gas is depicted as green rather than the usual pinkish-red. At right is an Earth-based view of the Eagle Nebula, in conventional colours.

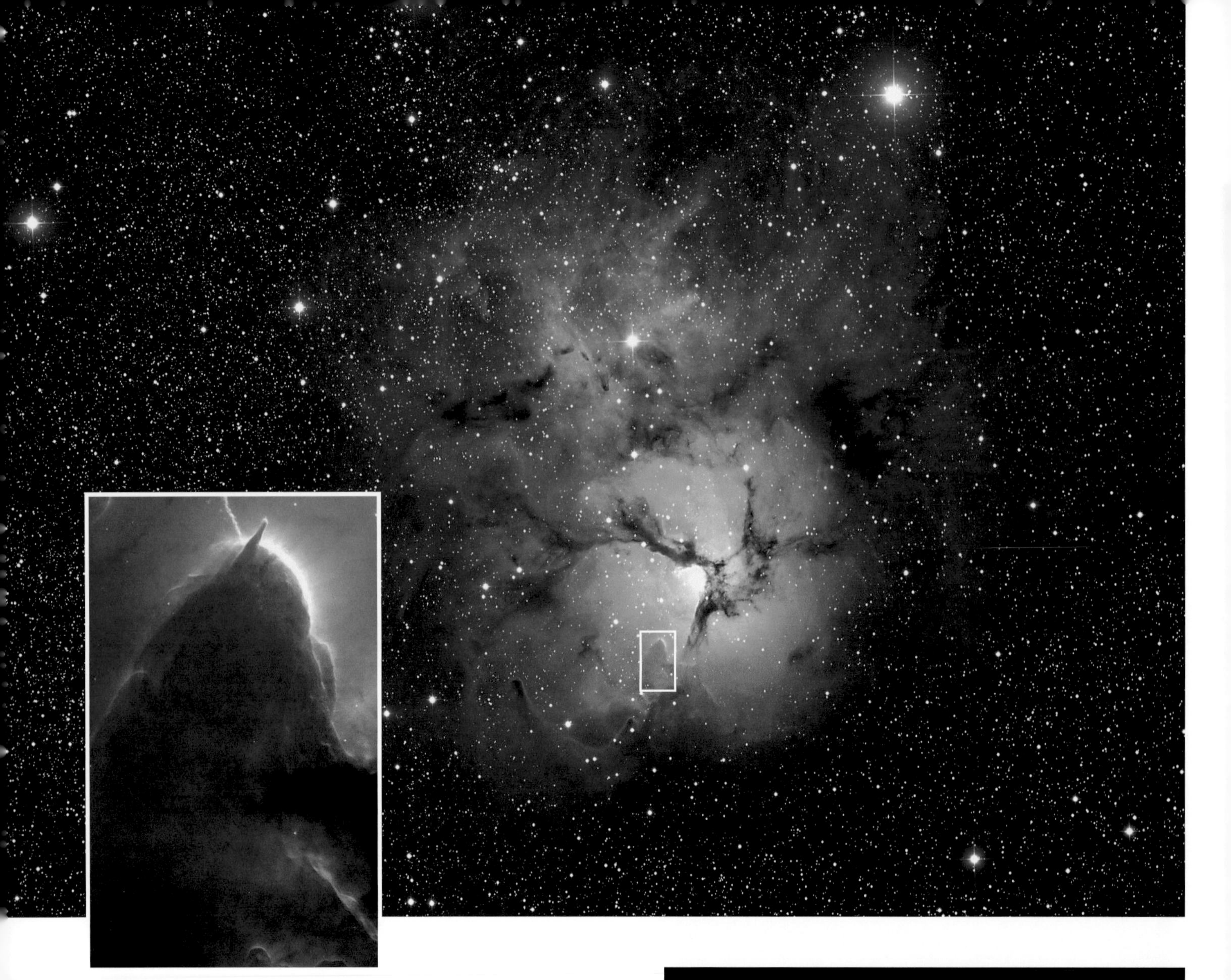

▲ **The Trifid:** Observers in the 19th century named this nebula the Trifid because it appeared to be trisected by three dark lanes; modern photographs, such as this one taken with the Anglo-Australian Telescope, show that there are in fact four lanes, which seem to wrap around the central cloud of glowing hydrogen gas. North of this is a separate haze of dust, illuminated by the bright star within; unlike the reddish hydrogen gas, this reflective dust cloud appears blue. The inset is the Hubble Space Telescope's view of the area outlined in white, showing a domical cloud of gas with a protruding finger at its summit like those in the Eagle Nebula (see previous page). The Trifid, which lies in Sagittarius, is about twice the size of the Orion Nebula.

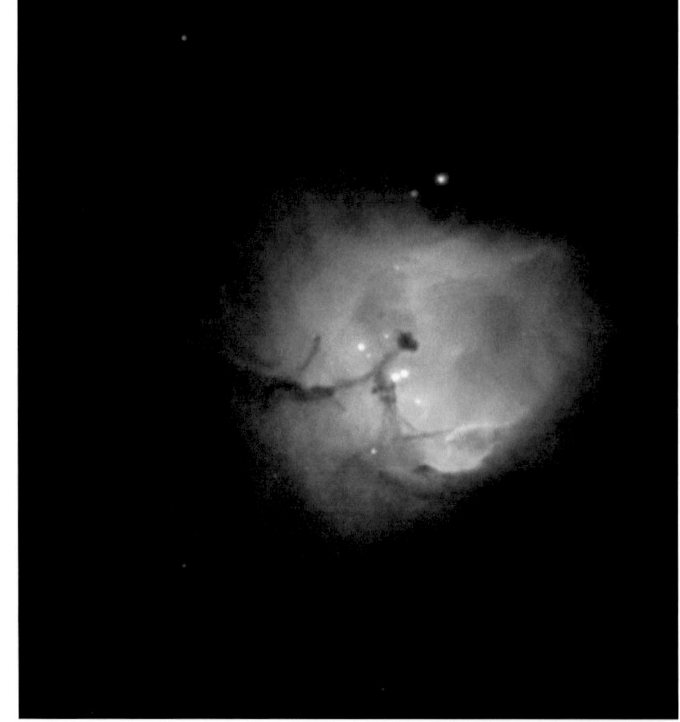

▶ **The Blob:** The nebula shown here, catalogued as N 81, lies in the Small Magellanic Cloud, a neighbour galaxy of ours some 200,000 light years away. The nebula is poorly seen from Earth, and astronomers had nicknamed it the Blob. Hubble's improved vision shows that N 81 has recently given birth to a family of massive stars, the two most luminous of which can be seen close together near its centre, each shining with the the brilliance of 300,000 Suns. Silhouetted against the brighter background is a wishbone-shaped veil of hydrogen molecules and dust from which future stars may emerge. The Small Magellanic Cloud has a simpler chemical composition than our own Galaxy, with scarcely any of the elements heavier than helium that are produced by multiple generations of stars. Hence nebulae such as N 81 provide an insight into stellar formation at a time when the Universe was still young.

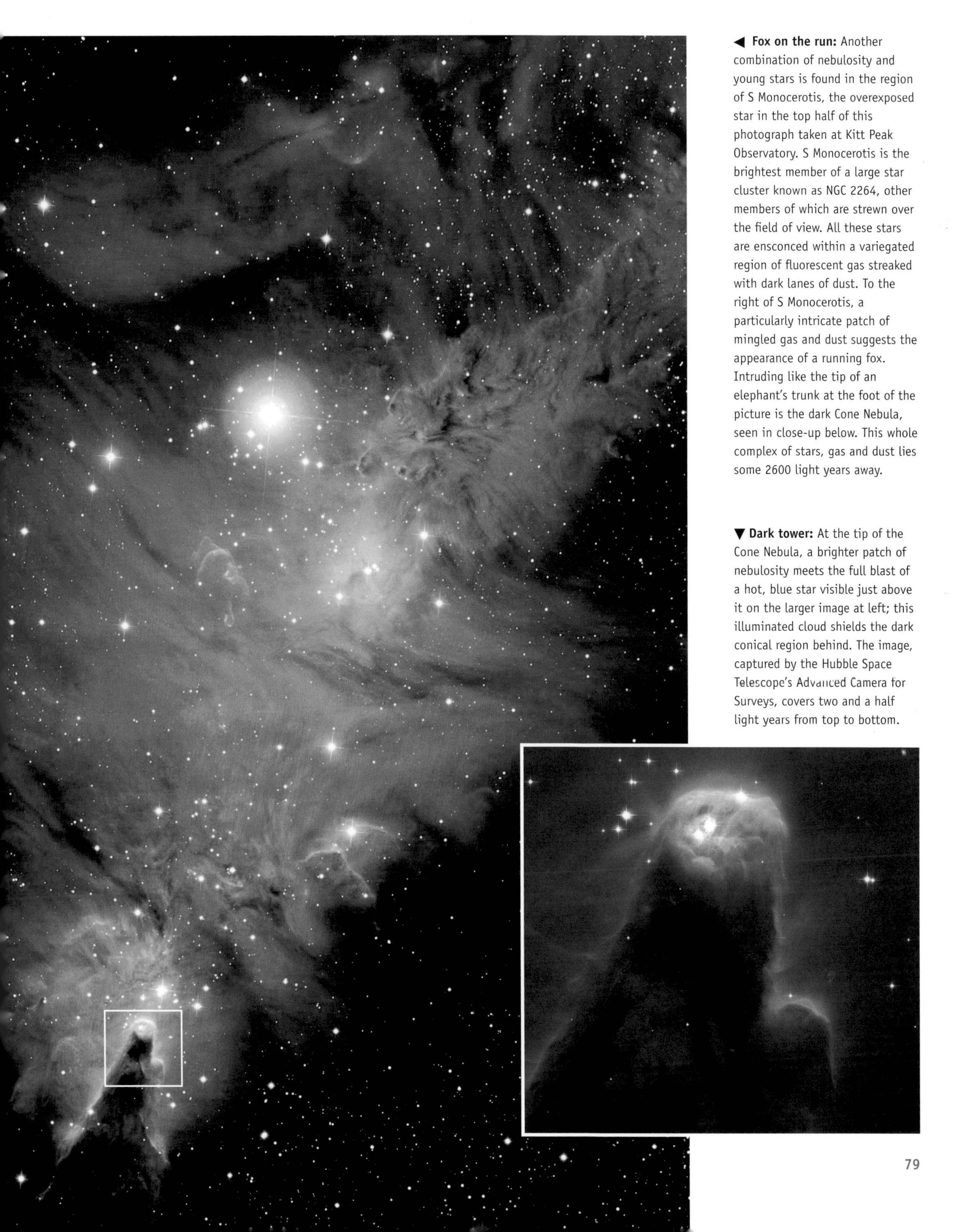

◀ **Fox on the run:** Another combination of nebulosity and young stars is found in the region of S Monocerotis, the overexposed star in the top half of this photograph taken at Kitt Peak Observatory. S Monocerotis is the brightest member of a large star cluster known as NGC 2264, other members of which are strewn over the field of view. All these stars are ensconced within a variegated region of fluorescent gas streaked with dark lanes of dust. To the right of S Monocerotis, a particularly intricate patch of mingled gas and dust suggests the appearance of a running fox. Intruding like the tip of an elephant's trunk at the foot of the picture is the dark Cone Nebula, seen in close-up below. This whole complex of stars, gas and dust lies some 2600 light years away.

▼ **Dark tower:** At the tip of the Cone Nebula, a brighter patch of nebulosity meets the full blast of a hot, blue star visible just above it on the larger image at left; this illuminated cloud shields the dark conical region behind. The image, captured by the Hubble Space Telescope's Advanced Camera for Surveys, covers two and a half light years from top to bottom.

STAR AND PLANET FORMATION

After more than a century of wondering whether planetary systems might exist around other stars, astronomers have now confirmed that they are a natural by-product of star birth and, in all probability, abundant throughout the Galaxy. New stars are ringed by a disk of surplus gas and dust, the tailings from their own construction. Planets grow within these by the collision and coalescence of dust particles. Building sites for new worlds, termed protoplanetary disks – proplyds for short – have been spied by the Hubble Space Telescope's keen eye around young stars, such as in the Orion Nebula (below). However, Hubble has also found that proplyds risk being evaporated by the radiation from nearby hot stars before planets have time to form, so perhaps many potential planetary systems are nipped in the bud.

Individual planets around other stars are too faint to see directly with current telescopes, but their presence can be inferred, notably through a gentle oscillation in a star's position as it and the unseen planet orbit their common centre of gravity. Powerful space telescopes currently under design may one day bring planets of other stars into direct view.

▲ **In at the birth:** Our own Solar System probably looked like this during its genesis 4.6 billion years ago. Here, the Hubble Space Telescope shows a protoplanetary disk of gas and dust orbiting a young star known as HH 30 in Taurus. The star is concealed by a horizontal dusty lane but its light reflects off the cloud above and below. In a cleaning-up process, gas falling onto the star from the encircling cloud is squirted out at high speed along the rotation axis (the vertical green jets). The white bar at lower left indicates a distance just over twice the diameter of Pluto's orbit.

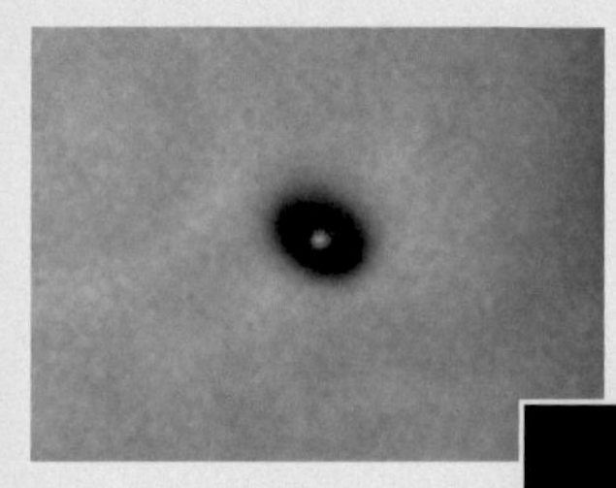

◀ **Diskworld:** A protoplanetary disk, silhouetted against the bright background of the Orion Nebula. The glow at the centre of the disk is a newly formed star about 1 million years old.

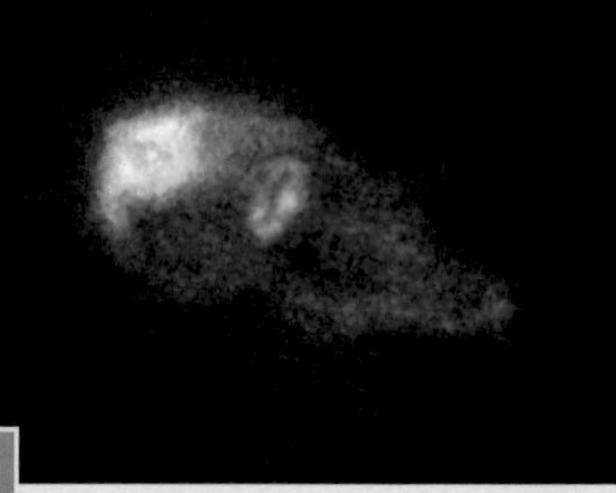

▶ **Being boiled:** This protoplanetary disk in the Orion Nebula, shown in green, is being evaporated by the energy of a nearby star, forming a tadpole-shaped cocoon.

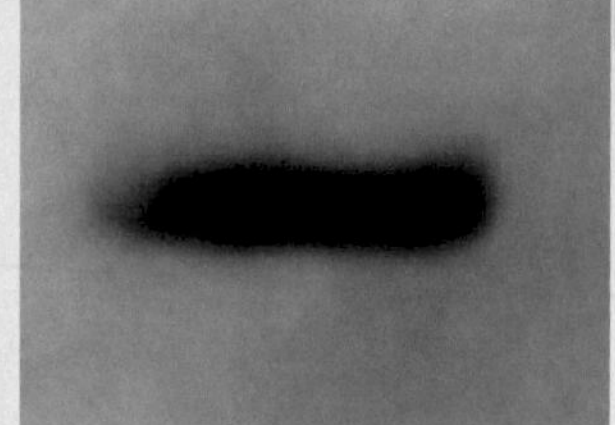

◀ **Edgewise:** A protoplanetary disk in the Orion Nebula seen edge on, obscuring the central star. Only 1% of the disk consists of dust, but that is enough to make it opaque.

▼ **Warp factor:** The star Beta Pictoris, 63 light years away, is orbited by a disk of dusty grains within which planets may already have formed. In this Hubble Space Telescope image, Beta Pictoris itself is blocked out by a black circle while the surrounding disk, which reflects the star's light, is depicted in false colour. The white inner disk is tilted by a few degrees from lower left to upper right with respect to the less-dense outer disk. Such a warping of the disk could be due to the gravitational effect of a Jupiter-sized planet orbiting in a dust-free zone closer in.

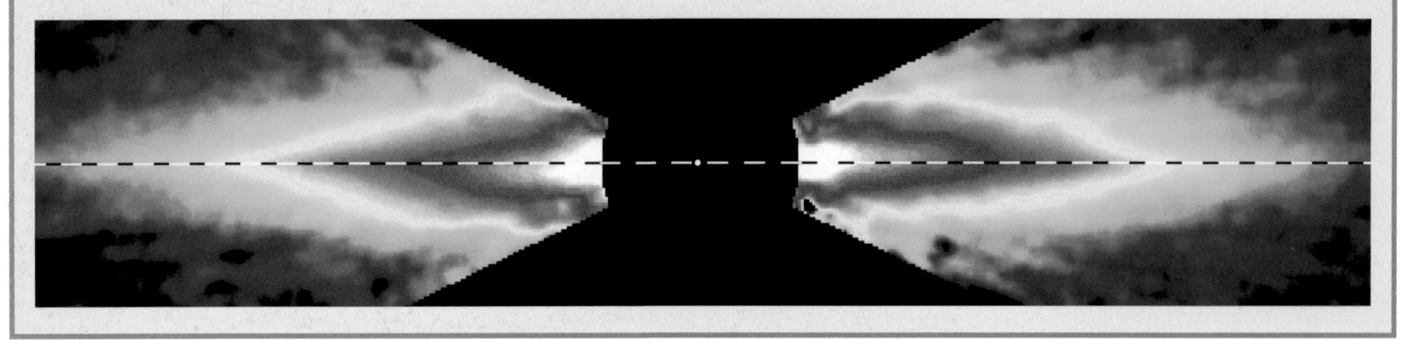

◀ ▼ **Through the Keyhole:** A view of a famous nebula by the ground-based Anglo–Australian Telescope, left, is compared with the superior vision of the Hubble Space Telescope, below. Called the Keyhole because of its distinctive shape, the nebula is a darker pond within the luminous expanses of the Eta Carinae Nebula. The Eta Carinae Nebula, visible as a brighter area in the southern Milky Way, is one of the largest gas clouds in the spiral arms of our own Galaxy, ten times wider than the Orion Nebula. The ground-based view portrays the full extent of the Keyhole, which is seen against the brightest part of the reddish Eta Carinae cloud; the area outlined in white is shown in detail below by the Hubble Space Telescope. The bulbous section, some 7 light years wide, is ringed by loops of hot, fluorescing gas along with tadpole-like blotches of cold gas and dust that will either evaporate or give birth to new stars, depending on their density. This Hubble view is presented in false colour to emphasize various features within the gas cloud. Among the massive young stars that have already formed within the bright nebula is Eta Carinae itself, seen to the lower left of the Keyhole in the ground-based view. Eta Carinae is enveloped in an elliptical cocoon of gas and dust that it threw off during an eruption in 1843. A Hubble close-up of Eta Carinae is on page 90.

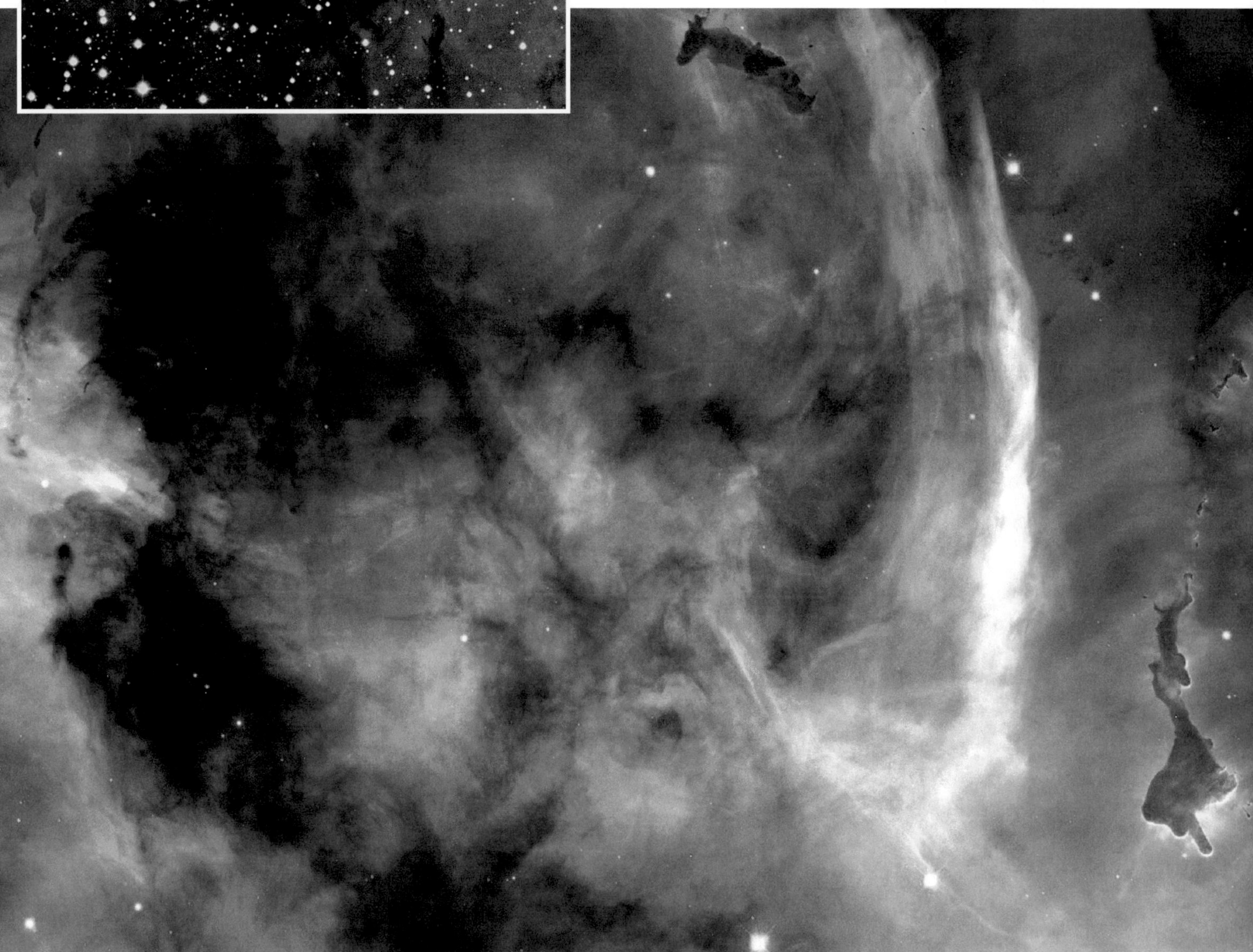

Star clusters

Large nebulae such as those illustrated on the preceding pages bring forth dozens or even hundreds of stars. Such youthful groupings of stars, termed open clusters, are found in the arms of spiral galaxies, where the gaseous nebulae that spawn them also reside. In most cases, open clusters will drift apart over time. Our own Sun was probably born in such a cluster, the other members of which have long since dispersed. The brightest open clusters can be seen with the naked eye – a notable example being the Pleiades, opposite – and hundreds are within range of binoculars and small telescopes. Because all the stars within a given cluster formed at approximately the same time from the same raw material, they provide a perfect natural laboratory for examining how the masses of different stars affect their luminosities and rate of evolution. Another type of cluster, the globular clusters, consist of dense concentrations of old stars bound together by gravity. They are not found in the spiral arms of galaxies but instead form halos around galaxies; a prime example is shown on page 98.

▼ **Doubling up:** These twin star clusters lie 7200 light years from us in a spiral arm of the Galaxy adjacent to the one in which the Sun resides. Known popularly as the Double Cluster, they can be seen with binoculars as two brighter clumps in the Milky Way in Perseus, marking the sword-wielding hand of the mythical hero. They are not identical twins, though. NGC 869, at right, is the more condensed while the most massive members of NGC 884, left, have evolved to become red giants and appear noticeably orange in this portrait taken by the Burrell Schmidt telescope at Kitt Peak observatory, Arizona.

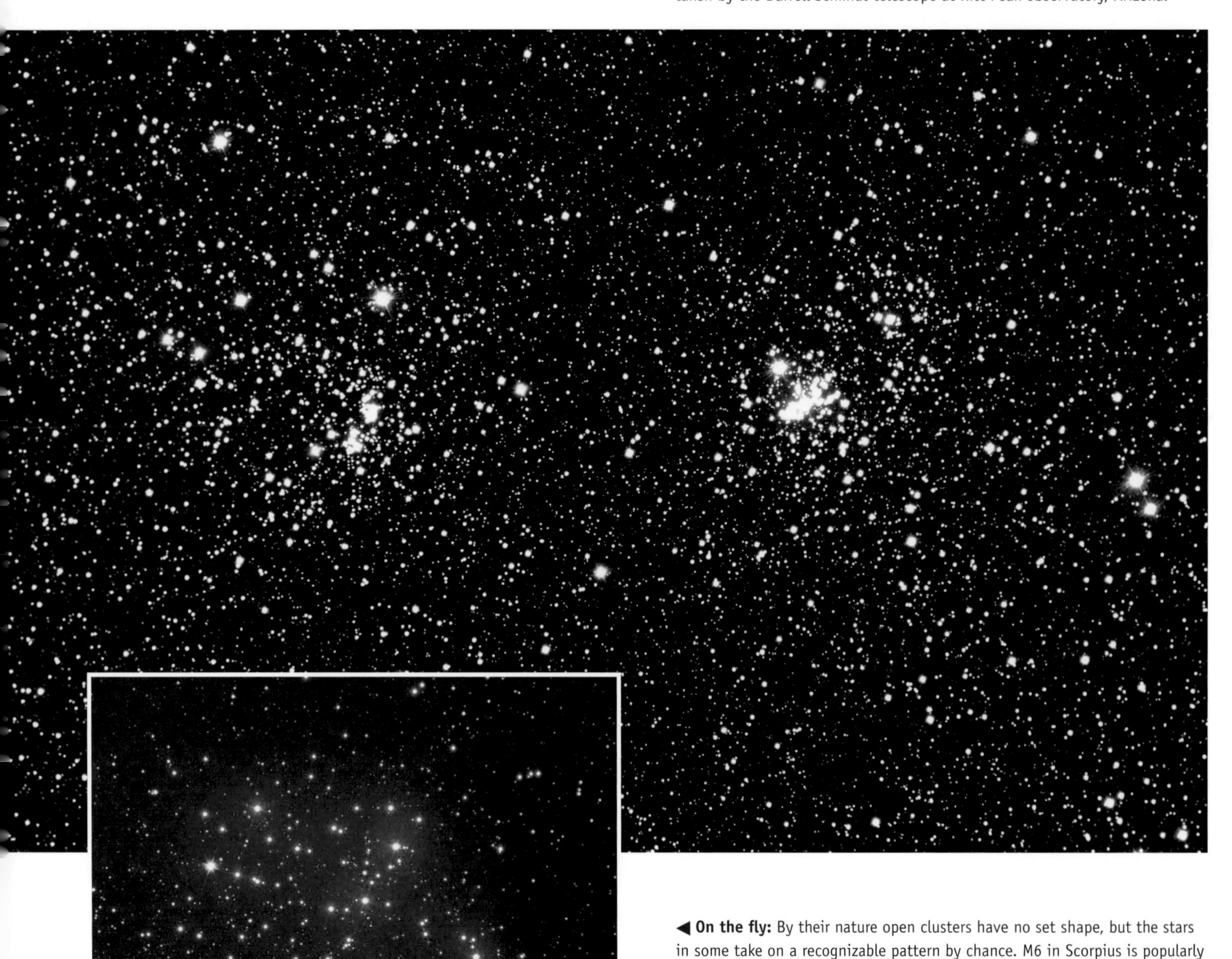

◀ **On the fly:** By their nature open clusters have no set shape, but the stars in some take on a recognizable pattern by chance. M6 in Scorpius is popularly known as the Butterfly Cluster, although there is also a strong resemblance to a bird in flight. Its most massive star has evolved into a red giant, at left.

SEVEN SISTERS – AND MORE

Hovering like a swarm of flies over the back of Taurus the bull is this sparkling star cluster, named the Pleiades after the seven sisters of Greek mythology. A handful of stars can be discerned with the naked eye, while binoculars and small telescopes bring dozens more into view. About 100 million years ago, when these stars were still forming from interstellar gas, the area now occupied by the Pleiades would have looked much like the Orion Nebula.

Enveloping the Pleiades is a wispy cloud of dust, well seen on the main photograph above taken with the United Kingdom Schmidt Telescope in Australia. This was once thought to be the remains of the nebular placenta from which the stars formed. Now, though, it is recognized to be an entirely separate cloud into which the stars have drifted by chance. The dust reflects the light of the brightest Pleiades stars and appears blue for the same reason that the sky does – blue light, being of short wavelength, is scattered more than red light. The inset, captured by the Hubble Space Telescope, is a close-up of an area close to the enshrouded star Merope, at the bottom of the cluster. Cascading shafts of light from Merope illuminate a denser part of the passing dust cloud. Merope's light decelerates the smaller particles by the phenomenon of radiation pressure, pushing them away like dust from a comet. Larger particles, less affected by radiation pressure, continue to move towards Merope.

▲ **Trigger mechanism:** NGC 1850 is an unusual open cluster in the Large Magellanic Cloud, a neighbouring galaxy some 170,000 light years away. The cluster is far more heavily populated than any open cluster in the Milky Way, containing around 6000 stars. NGC 1850 is estimated to be about 50 million years old, but next to it lies a smaller cluster less than one-tenth its age – the pocket of bright starlight to the lower right of this Hubble Space Telescope image. Early in the history of the main cluster the most massive stars within it exploded as supernovae, creating the curtains of glowing gas shown in blue at the left of the image. Blast waves from the supernovae are thought to have triggered a new generation of star formation in nearby clouds of gas, giving rise to the younger cluster. In time, the most massive stars in this younger cluster will themselves explode, possibly setting off another round of star formation nearby.

▶ **Life story:** Multiple stages in the life cycle of stars are captured in this Hubble Space Telescope portrait of the area around the open cluster NGC 3603 in the southern constellation Carina. At the upper right of the frame are dark globes of cold, dense gas, sometimes known as Bok globules after the Dutch-American astronomer Bart Bok, which are thought to be an early stage in the collapse of an interstellar cloud on its way to becoming a star. In the lower half are gaseous pillars being eroded by starlight, similar to those in the Eagle Nebula (page 77) and the Cone Nebula (page 79). NGC 3603 itself, at left of centre, is a young cluster containing extremely hot and massive stars which will one day explode as supernovae. To the lower left of the cluster are two bright, tadpole-shaped nebulae similar in appearance to the so-called proplyds around young stars in the Orion Nebula shown on page 80. However, these proplyds are five to ten times larger in size, and correspondingly greater in mass, than those in Orion. Finally, to the upper left of the cluster is a blue supergiant star nearing the end of its life, encircled by a ring of gas thrown off from its equator. Two additional blobs of gas – one to its upper right and one to its lower left – have been ejected from the star's poles. The grey-blue colour of this ejected gas indicates that it has been chemically enriched by nuclear processes in the aged star. This Hubble view is presented in natural colour.

◀ **Ghostly remains:** NGC 3293 is a gathering of starlight to the north of the great Eta Carinae Nebula, over 8000 light years from us. The brightest stars in this cluster are hot and hence blue – with one exception. As in many young clusters, the most massive member has evolved into a red giant and provides a noticeable colour contrast. The nebulosity from which clusters form is usually rapidly dispersed by the energy of the exuberant young stars within it, but in this case some ghostly traces remain. The reflective dust appears blue, while the hydrogen gas glows a characteristic red on this photograph taken with the Anglo-Australian Telescope.

Star death

Like the engine of a car, a star's power output affects its fuel consumption. High-performance stars have the highest consumption and hence their tanks run dry the quickest – in a matter of a few million years. Low-output stars are the most economical of all, lasting tens of billions of years. The Sun, as we have seen, is a mid-range model with modest power but good fuel consumption. Even so, if we were to come back in a few billion years' time, we would see signs of age caused by the spread of hydrogen burning outwards from its core: the Sun would be swelling in size and its surface temperature would be dropping so that it appeared redder than it does today. The Sun would have become a red giant, dozens of times larger than at present and hundreds of times more luminous.

By then, the temperature of Earth will have long since exceeded the limits endurable by the hardiest forms of life. With its seas boiled into steam, the Earth will resemble Venus. Eventually, the red giant Sun may swell enough to engulf the Earth, leaving our former home planet a charred cinder. Humans – or their distant descendants – will have long since fled to the haven of a younger star.

At its largest, the senescent Sun will undergo a glorious metamorphosis. Too distended to be held in check by the Sun's gravity, its outer layers will slough off into space, garlanding the dying star with loops and shells of gas. Such objects were given the name planetary nebulae by 18th-century observers who thought their gaseous disks resembled distant planets. Many are now known that do not look like planets at all, but the name has stuck.

Stripped bare by the expulsion of the overlying gas, the white-hot core of the former star is exposed to view as a so-called white dwarf, little larger than the Earth yet containing about as much mass as the Sun. Ultraviolet light from this hot dwarf ionizes the surrounding gas, which glows with a range of colours depending on its temperature and composition.

Nuclear reactions have ceased in white dwarfs. Since they are no longer generating energy, there is nothing to prevent the inward crush of gravity from packing the atomic particles within them as closely together as is physically possible. As a result, white dwarfs are denser than any material on Earth – a spoonful of matter from one would weigh as much as a truck. Over billions of years, the white dwarf cools and fades to invisibility.

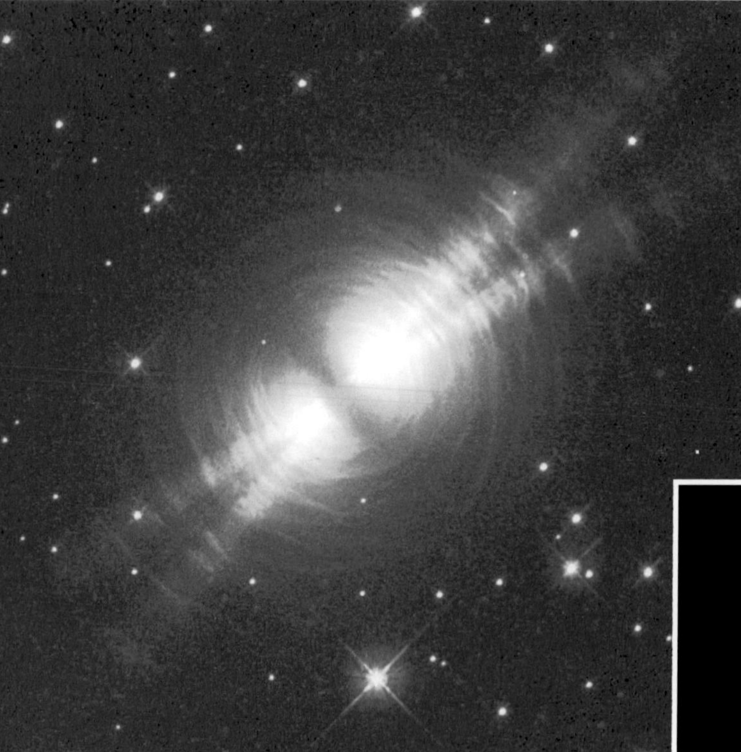

▲ **Hatching out:** The Egg Nebula in Cygnus appears as though seen through frosted glass in this false-colour view from the Hubble Space Telescope. The frostings are actually shells of carbon dust ejected by the former red giant star at intervals of 100 to 500 years. Most of the carbon in the Universe, on which life is based, is manufactured and dispersed back into space in this way by red giants. A dark ring of dust around its equator obscures the central star from our view, but shafts of light from the star's poles project through the obscuration and reflect off the surrounding dust.

▼ **Eyeful:** NGC 6543, nicknamed the Cat's Eye, is one of the most complex planetary nebulae known. Its features, captured in detail by the Hubble Space Telescope, suggest that the central star is a close double. On this interpretation, the existence of the companion has helped shape the ejection of shells of gas from the central star. Some of the gas may have been collected by the companion and propelled along its rotation axis to produce the spiralling jets of gas, shown red, at each end of the lobes.

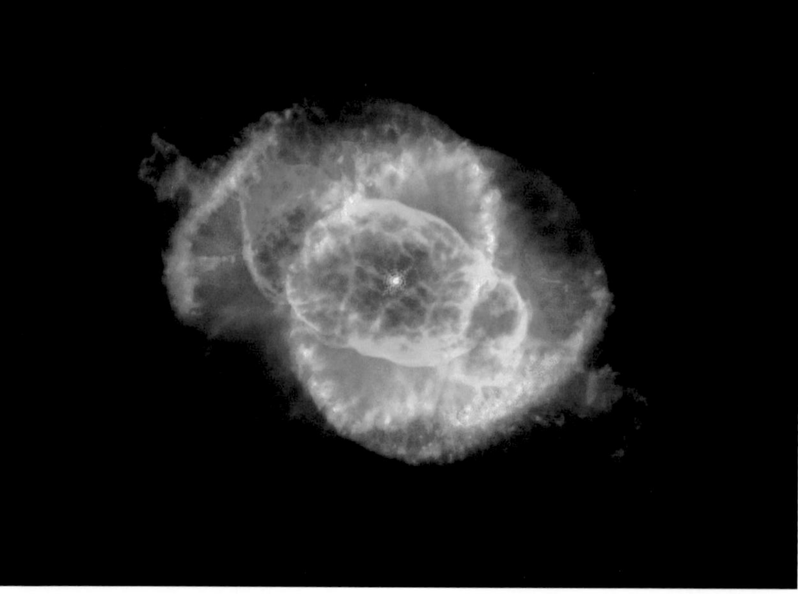

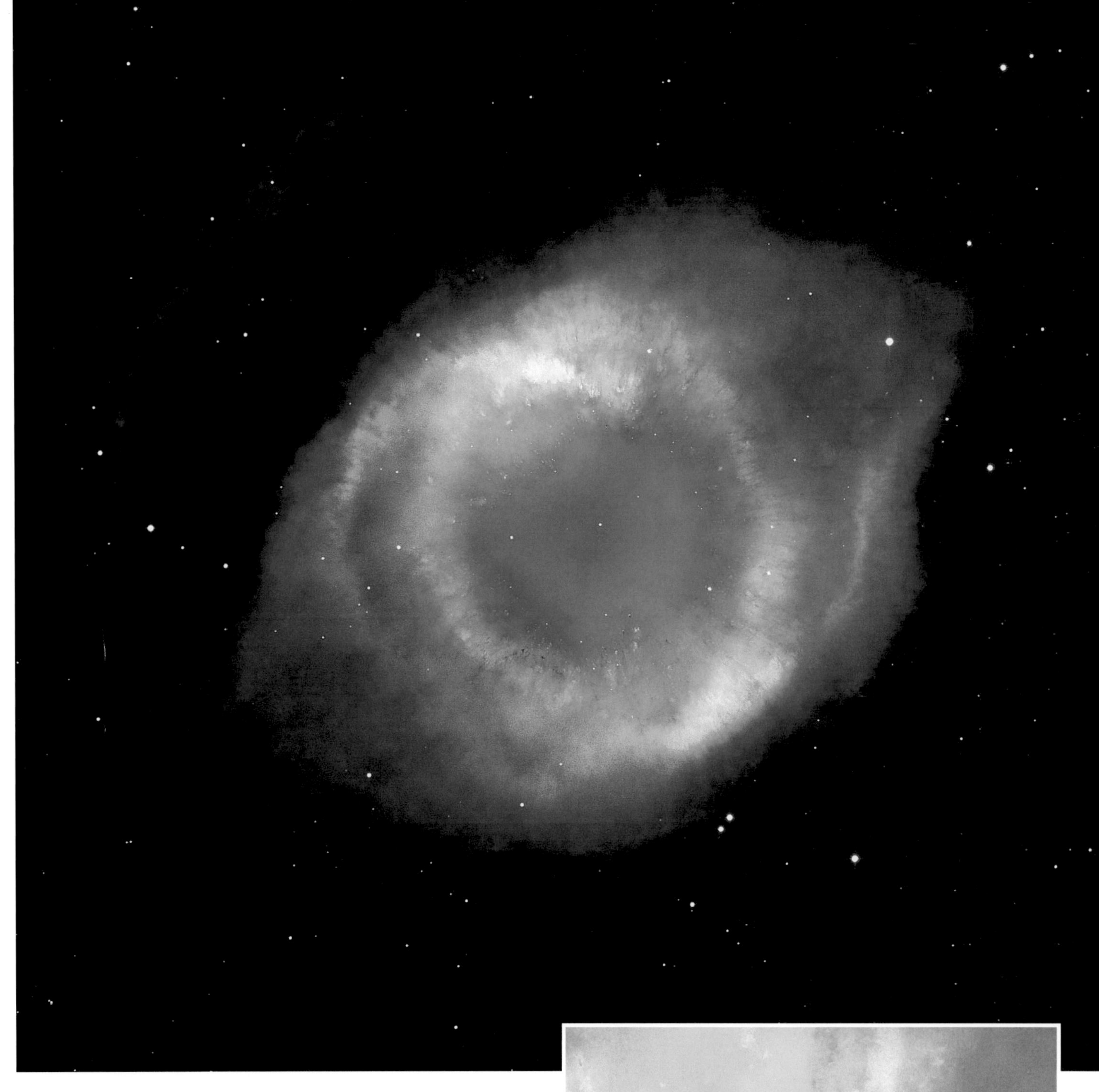

▲ **Simple elegance:** One of the most symmetrically elegant planetary nebulae is the Helix in the constellation Aquarius, seen here in a stunning composite of narrow-angle images from the Hubble Space Telescope's Advanced Camera for Surveys blended with a wide-angle view from the 0.9-m telescope at Kitt Peak National Observatory, Arizona. The exposed core of the star that ejected the gas is easily identified at the centre. Oxygen atoms glow blue in the inner part of the nebula, excited to fluorescence by the star's ultraviolet light, while the outer region gets its red coloration from nitrogen and hydrogen. The Helix is the closest planetary nebula to us, 300 or so light years away.

▶ **In the swim:** On the inner periphery of the Helix's ring, the Hubble Space Telescope's Advanced Camera for Surveys netted these tadpole-like clumps, each at least twice the diameter of our Solar System, swimming in a blue pool of oxygen gas. The 'tadpoles' are thought to arise where hot, low-density gas streaming from the central white dwarf collides with cooler, denser gas and dust that was ejected by the doomed star around 10,000 years ago.

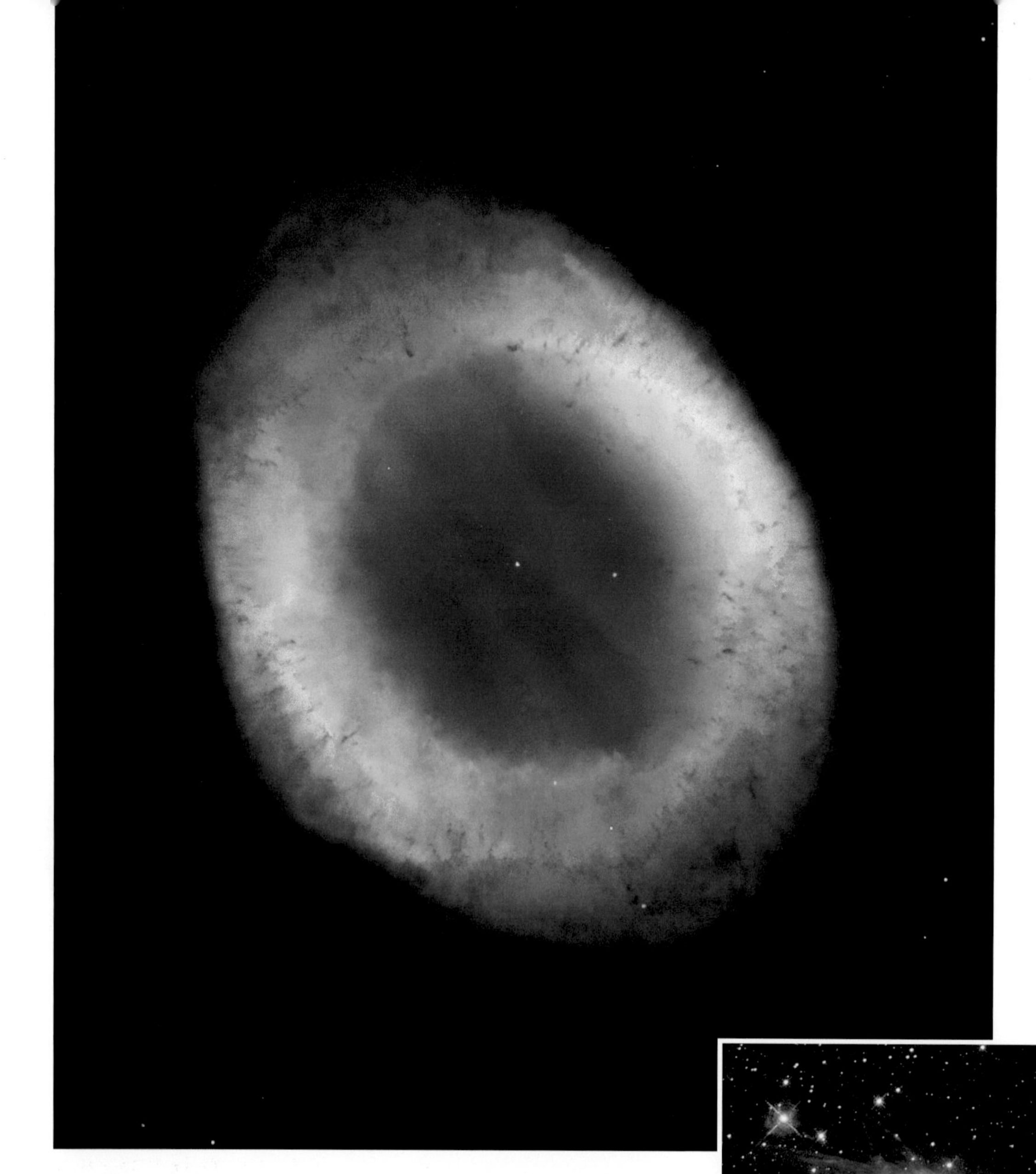

◀ **Ringing the changes:** The confusing variety of shapes exhibited by planetary nebulae can be explained in part by the different angles at which we view them. For example, the famous Ring Nebula in Lyra appears to be a simple shell of gas encircling the dying central star, as shown in this true-colour photograph from the Hubble Space Telescope. In fact, astronomers now realize that most planetary nebulae consist of two lobes of matter streaming from the poles of the central star, as shown in the photograph below. In the case of the Ring Nebula and those of similar shape, these lobes are seen end-on. Where the nebula is seen at an intermediate angle so that the lobes overlap, more complex shapes such as that of the Cat's Eye (page 86) may result.

▲ **Antics:** The Ant Nebula is named for its insect-like appearance, caused by two lobes of gas ejected by a dying red giant star. Here, the Hubble Space Telescope views them side-on; but if seen end-on, the lobes would probably resemble a ring structure like the other planetary nebulae pictured here.

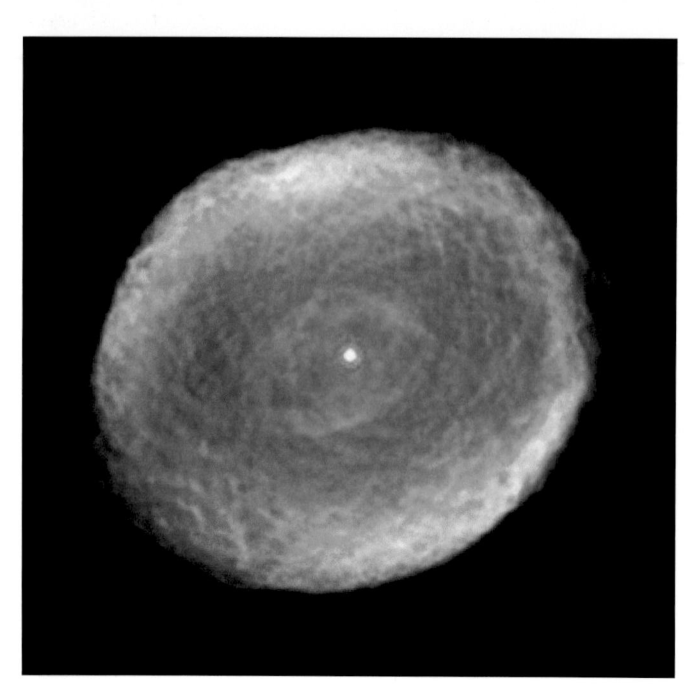

◀ **Spirograph:** Filigree patterns within the planetary nebula IC 418 in the constellation Lepus give rise to its popular name of the Spirograph. Their cause is not known. In this Hubble Space Telescope picture, red represents emission from nitrogen (the coolest gas in the nebula, located farthest from the hot nucleus), green shows emission from hydrogen, and blue indicates the emission from oxygen (the hottest gas, closest to the central star).

▶ **Fringe benefit:** Aided by the stunning clarity of the Hubble Space Telescope's images, astronomers can now interpret many of the features of planetary nebulae which once were enigmatic. The Eskimo Nebula in Gemini gets its name from its supposed resemblance to a face fringed by a fur parka. Astronomers now recognize that it consists of two elliptically shaped lobes of gas seen end-on. These lobes expand from the central star's poles; they are held back around the equator by a ring of denser gas and dust that was thrown off in the last gasps of its red giant phase. Each lobe is about 1 light year long and half a light year wide. As in the Helix Nebula (page 87), the comet-shaped features in the 'parka' may be due to a collision of fast- and slow-moving gases. Nitrogen gas in the nebula is shown red, hydrogen is green, oxygen blue and helium violet.

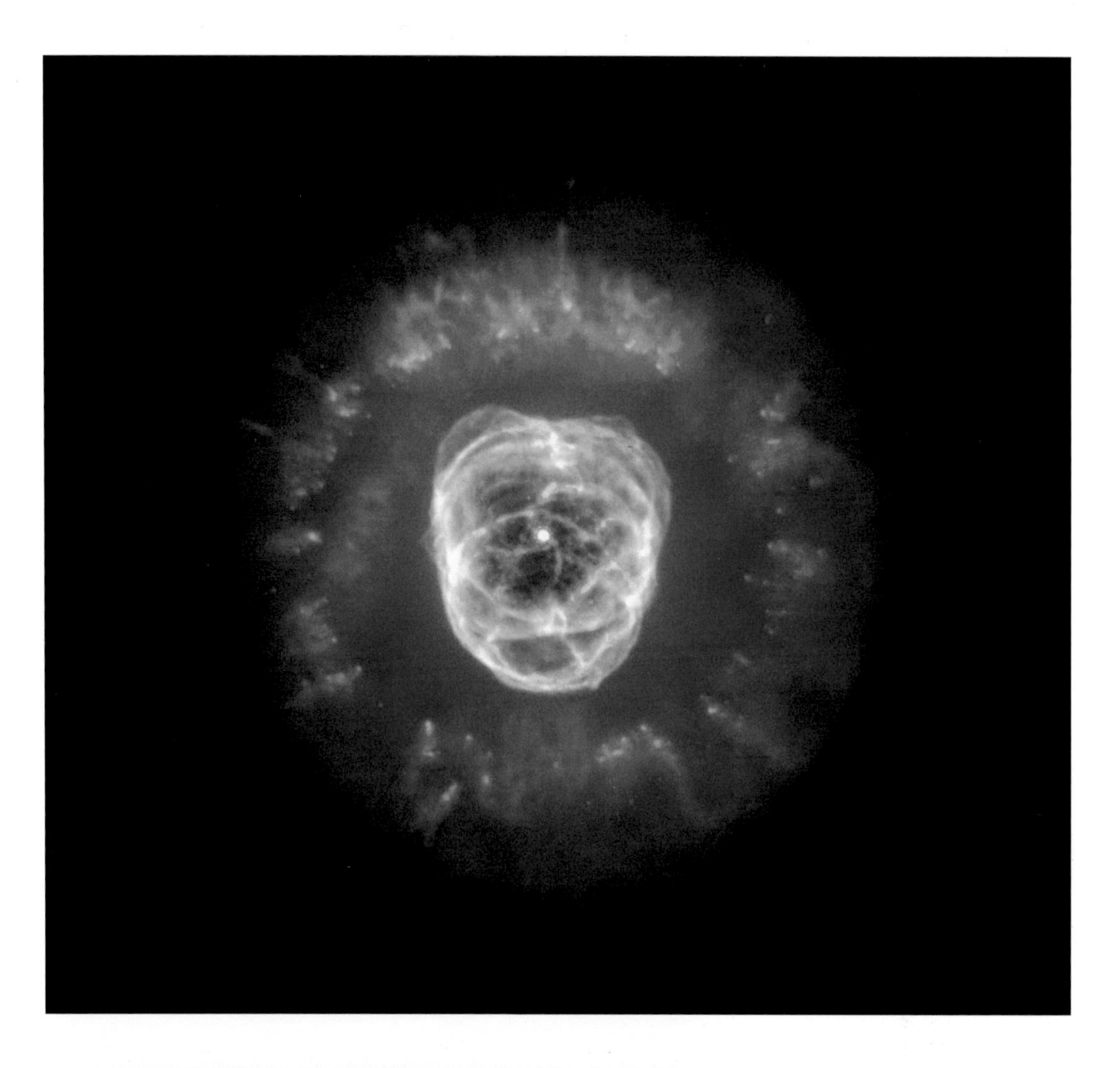

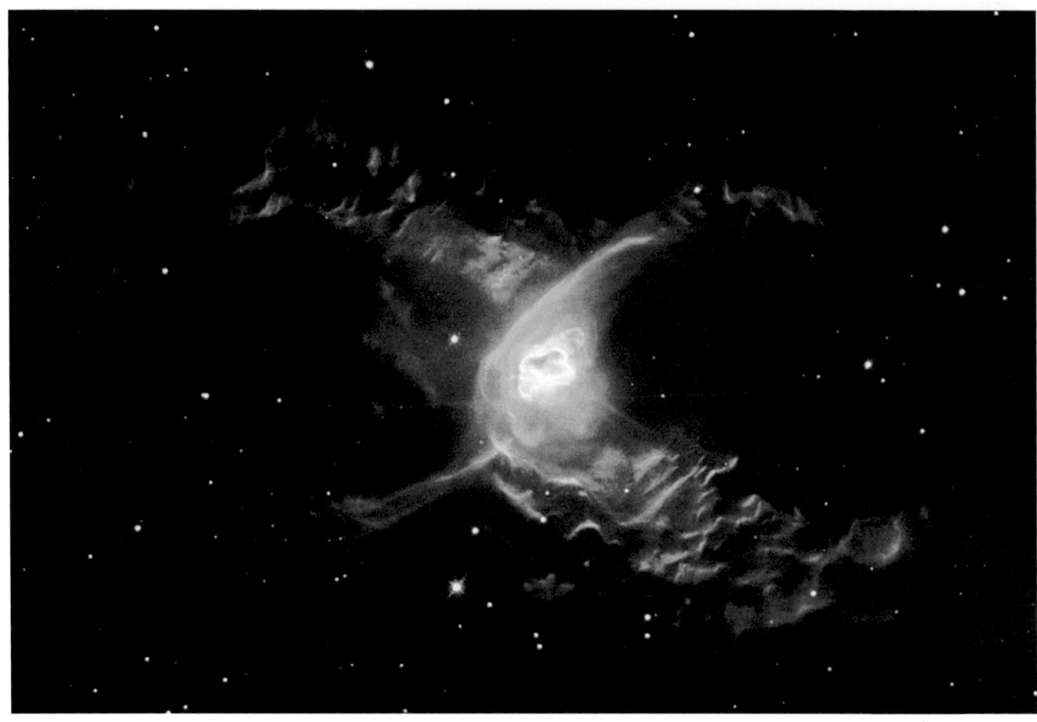

◀ **Red alert:** Adding to the almost limitless variety of forms displayed by planetary nebulae is the Red Spider Nebula in Sagittarius. At its centre is one of the hottest stars known, with a temperature of at least 500,000 degrees. High-speed streams of atomic particles from this star, known as stellar 'winds', ripple the surface of the surrounding gas lobes, generating flowing waves 100 billion km (60 billion miles) high – enough to satisfy the most advanced interstellar surfers.

VIOLENT DEATH OF A HEAVYWEIGHT

Stars much more massive than the Sun undergo an altogether more violent demise. Burning their fuel at a turbocharged rate, these dazzling superstars race through their lives in only a fraction of the time taken by those like the Sun. In their advanced years they distend not merely into red giants but supergiants, hundreds of times larger than the present-day Sun and at least 10,000 times brighter. But this is only the prelude to one of the most cataclysmic events in the Universe: a supernova explosion.

A supernova occurs once a supergiant has used up the last of the energy resources at its core and its central fires have shut down. With nothing left to shore it up against the immense overburden of the layers above, the star collapses in on itself, hammering together the electrons and protons at its centre to produce a ball of neutrons that resists further compression. The infalling gas rebounds violently off this incompressible core, creating a shock wave that triggers a chain of nuclear reactions in the star's outer layers and hurls them into space at speeds up to 70 million km/h (45 million mile/h). All the chemical elements of nature are forged in the thermonuclear cauldron of a supernova. Scattered into space by the blast, they mix with the interstellar gas of nebulae, later to be collected up into new stars and planets, completing a cycle of stellar life and death. We all contain atoms that were created in the supernova explosions of massive stars that lived and died long before the Sun was born.

So much energy is released in a supernova outburst that it can blaze as brightly as all the stars in a small galaxy for weeks on end. Supernovae in other galaxies are regularly seen through telescopes, but none have been spotted in our own Galaxy since 1604. Our best view of a supernova in recent times came in 1987 when one erupted in a nearby galaxy, the Large Magellanic Cloud, remaining visible to the naked eye for almost a year (see facing page). Were one to go off within a few hundred light years of us, it would blaze as brightly in our skies as the Moon.

As at any demolition site, the wreckage from a supernova is strewn around for all to see. Loops of glowing gas decorate the starfields of the Milky Way where supernovae have erupted in times past. Most celebrated of these is the Crab Nebula, the remains of a supernova seen from Earth as a temporary 'new' star in Taurus in AD 1054 (page 93). Shreds of prehistoric supernovae can be found draped across the sky in Cygnus and Vela.

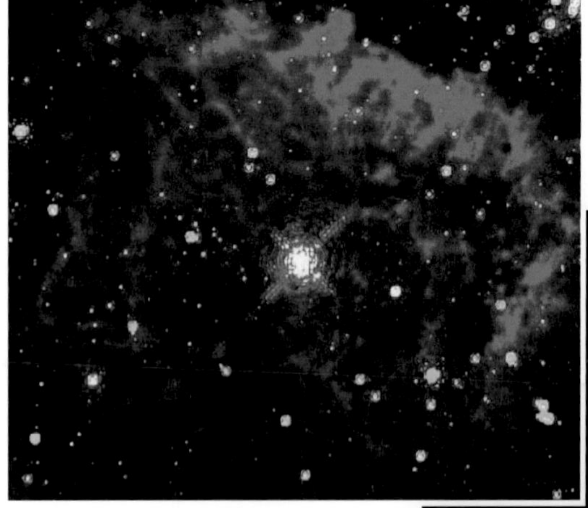

▲ **Smoking gun:** The so-called "pistol star" illustrates what happens to a star born with a serious weight problem: it sheds the excess mass in a series of eruptions. Over the past few thousand years it has ejected several solar masses of gas, forming a pistol-shaped nebula visible in this false-colour image from the Hubble Space Telescope. Despite losing so much gas it is still one of the biggest and most powerful stars known, and will end its life in a supernova explosion.

▼ **Supernova in the making:** Eta Carinae is a hot, massive star on its way to becoming a supernova. Unstable because of its intense luminosity, it flared up in 1843 when it was temporarily the second-brightest star in the sky. During that outburst it threw off an elongated cocoon of gas and dust, known as the Homunculus because of its humanoid shape, which now shrouds the star from view. The Homunculus is seen in detail in this Hubble Space Telescope image.

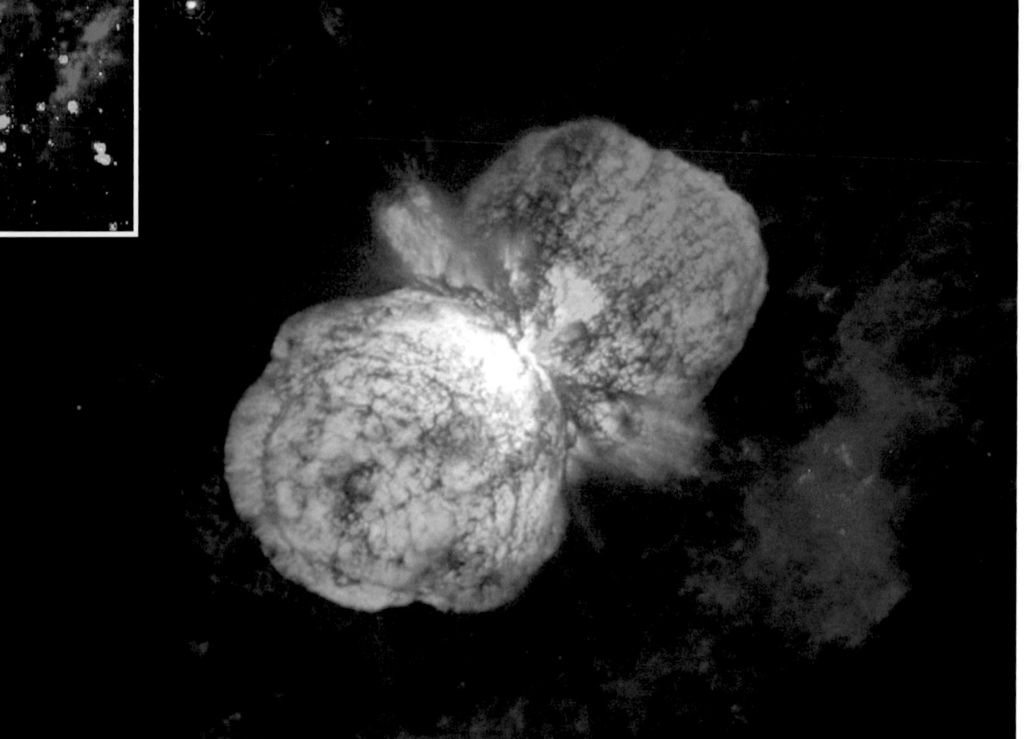

SUPERNOVA 1987A

In February 1987 a massive star died violently in the Large Magellanic Cloud, a satellite galaxy of ours 170,000 light years away in the southern sky. Supernova 1987A, as it was called, was the nearest supernova for nearly 400 years and the only one in all that time bright enough to be visible to the naked eye. The pair of photographs at upper right, taken by the Anglo-Australian Telescope, show the star before it exploded (arrowed) and two weeks afterwards. The main picture at right is a view from the Hubble Space Telescope seven years later, showing the supernova at the centre of a bright inner ring; this ring appears elliptical because it is seen at an angle. It consists of gas thrown off the star's equator during its supergiant phase thousands of years before it erupted and is now lit up by the energy of the supernova. Two other overlapping rings, one in front of it and the other behind, were also thrown off around that time, from the star's poles. Two brighter foreground stars, unconnected with the supernova, appear superimposed on the rings by chance.

NEUTRON STARS, PULSARS AND BLACK HOLES

What of the central core created in the supernova explosion? It remains as a so-called neutron star, a thousand times smaller and a million times denser than the white dwarfs left behind by the death of stars like the Sun. A typical neutron star contains the mass of one or two Suns, compacted into a ball only about 25 km (15 miles) across, the size of a city. The resulting density is so great that a spoonful of matter from one would weigh as much as several thousand fully laden oil tankers.

Being so small, neutron stars can spin rapidly, once every few seconds or faster. A beam of radiation, channelled outwards from their magnetic poles, sweeps around the sky as they spin. If this beam intercepts the Earth we observe the neutron star as a rapidly flickering radio source known as a pulsar. Some young and energetic pulsars, notably the one in the Crab Nebula, have also been seen blinking at optical wavelengths, in time with the radio pulses.

Should the core created in the collapse of a supergiant have a mass greater than about three Suns, an even more astonishing fate awaits. So strong is the gravitational pull of such a small yet massive object that it collapses beyond the stage of a neutron star. It shrinks ever smaller and denser, disappearing into the oblivion of a black hole – a region of space where gravity's clutches are so strong that not even the star's own light can escape.

At the black hole's centre, the remains of the former star are crushed to a point of infinite density, termed a singularity. The dying star has, in effect, squeezed itself out of existence. By their nature, black holes are invisible, but they give their presence away by the effects they have on the stars and gas around them. For example, black holes can swallow gas from their surroundings. As the gas spirals in towards the black hole it heats up to many millions of degrees. At such temperatures the gas emits X-rays that can be detected by satellite observatories and in this way the existence of numerous black holes has been deduced.

▶ **Analysing a supernova's composition:** X-ray emission from hot gas expelled by supernova explosions allows astronomers to identify the chemical elements present in the debris. Here, the supernova remnant Cassiopeia A is seen by the Chandra X-ray observatory satellite. Gas colour-coded red on the left outer edge is enriched in iron that was created by nuclear reactions deep within the star at temperatures of four to five billion degrees; the bright greenish-white region on the lower left contains silicon and sulphur formed in surrounding layers at three billion degrees. From the rate of expansion of the gas shell, the supernova that created Cassiopeia A is estimated to have exploded around 1680 but did not reach naked-eye brightness from Earth.

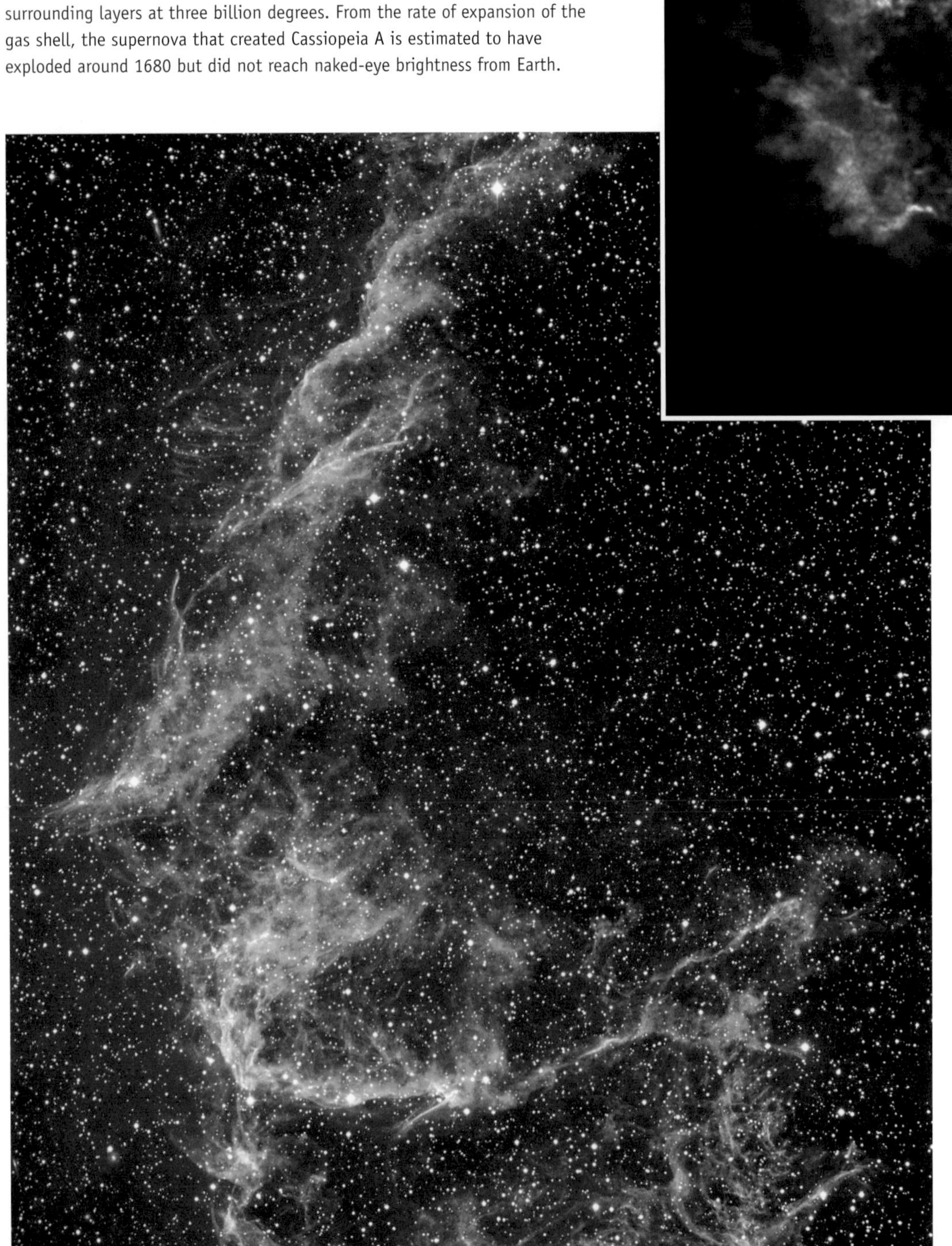

◀ **Star-spangled banner:** Colourful ribbons of gas draped in great swags among the stars in Cygnus are the remains of a star that exploded as a supernova in prehistoric times. Popularly known as the Veil Nebula, this is the brightest segment of a bubble of gas 75 light years across. The nebula is shown here in a true-colour photograph from the UK's Isaac Newton Telescope in the Canary Islands. According to recent measurements of the nebula's expansion rate, the supernova that formed it must have exploded about 5000 years ago. At an estimated distance of 1500 light years, it would have been bright enough to cast shadows on Earth.

THE CRAB NEBULA

In the constellation Taurus lies one of the most celebrated objects in the heavens: the shattered remains of a supernova that was seen by observers on Earth in AD 1054. At its best it was said to have outshone the planet Venus and was visible to the naked eye for over a year in all. Wreckage from the explosion is still visible to small telescopes nearly 1000 years later, expanding at about 1500 km/s (900 mile/s). The remnant is popularly known as the Crab Nebula, from a description by the 19th-century Irish astronomer Lord Rosse who likened it to a hermit crab. Another designation is M1, the first in a list of objects that might be confused with comets that was drawn up in the 18th century by the Frenchman Charles Messier.

Seen above in a portrait taken by the 5-m (200-inch) Palomar reflector, the Crab Nebula consists of a milky centre surrounded by reddish fronds of hydrogen gas. Near the heart of the nebula are two stars, the lower of which is the superdense core of the exploded star (see inset). This remnant core is now a pulsar, emitting radio and optical flashes as it spins 30 times a second. Energy from the pulsar keeps the nebula glowing. The Crab Nebula is 6500 light years away and roughly 10 light years in diameter.

▲ **Peering at a pulsar:** Zooming in to the heart of the Crab Nebula (the boxed area on the main picture above), the Hubble Space Telescope shows the pulsar created in the supernova explosion, left of centre; the star above it is an unrelated foreground object. The pulsar heats its surroundings, creating a ghostly bluish-green gas cloud, including a blue arc just to its right. In this false-colour view, yellowish green filaments of gas at the bottom lie in front of the pulsar and are approaching us, while the orange and pink filaments include receding material beyond the pulsar.

Galaxies and the Universe

Pink clouds of gas blossom along the curling arms of NGC 2997, a classic spiral galaxy about 40 million light years away in the southern constellation Antlia, portrayed here by the Anglo–Australian Telescope.

Every star we see in the night sky belongs to a vast congregation called the Galaxy. Its more distant members mass into the misty band of the Milky Way, a name sometimes also applied to the Galaxy itself. Seen from outside, our Galaxy would appear as a spiral, like a Catherine wheel, over ten times wider than it is thick. Like a city, our Galaxy has a crowded centre and more spacious suburbs. The Sun is a suburban star, residing quietly in a spiral arm some 25,000 light years from the bustling city centre.

Until the 1920s it was not known whether anything lay beyond the Milky Way other than empty space. Using the newly opened 2.5-metre (100-inch) reflector on Mount Wilson in California, the American astronomer Edwin Hubble established that various misty-looking patches of light, long known to astronomers but of undefined nature, were in fact separate galaxies. Most prominent of these was the great nebula visible to the naked eye in Andromeda. When the wan light of this object was focused onto sensitive photographic plates by the Mount Wilson telescope, Hubble could at last see that it consisted of faint, distant stars.

Distances between the stars in our Galaxy seem immense, but distances between galaxies are a million times larger. According to modern measurements, the Andromeda Galaxy (no longer classified as a nebula) lies some two and a half million light years away, yet it is a close neighbour on the intergalactic scale. The Andromeda Galaxy and our own Galaxy are the two largest members of a modest cluster of a few dozen members known as the Local Group. It turns out that most galaxies are members of clusters, some far larger than the Local Group.

While pursuing his galactic prey to ever greater distances Hubble made his most astounding discovery, which underpins our present understanding of the Universe: galaxies (or, more accurately, clusters of galaxies) are moving apart from each other as the space between them expands. Most astronomers now agree that the entire Universe must once have been crammed together in a state of unimaginably high density and temperature. This scramble of matter and energy was flung outwards by a cataclysmic explosion which marked the origin of the Universe as we know it. That explosion is termed the Big Bang.

By measuring the expansion rate we can deduce how long must have elapsed since the Big Bang. This work, begun by Hubble and refined by astronomers using the orbiting telescope named after him, yields an age of around 13 billion years, about three times older than the Earth and consistent with the ages of the oldest known stars. As far as astronomers can tell, the Universe will continue to expand forever, thinning out and eventually fading into darkness as the last stars burn out.

A TOUR OF THE MILKY WAY

This 360-degree panorama captures the entire stardust band of the Milky Way as seen from Earth, along with other familiar landmarks in the night sky. In the bottom left corner is the Pleiades star cluster, hazy and bluish. To its upper right lies the reddish California Nebula, faint visually but prominent photographically. Above it, on the northern fringe of the Milky Way, is the bright star Capella. About 5½ cm (2¼ inches) to the right of the Pleiades lies the silvery streak of the Andromeda Galaxy. Near the top of the image sits Polaris, the north pole star. Another 4 cm (1½ inches) along the plane of the Milky Way is the pink, hook-shaped North America Nebula, alongside the bright star Deneb in the constellation Cygnus. Above right of Deneb is brilliant blue-white Vega. Stretching away from Deneb is the Great Rift, a dark horse's tail of dust in our local spiral arm of the Galaxy which widens through perspective as it sweeps towards us out of the stellar background, petering out above the yolky galactic centre. Upper right of the Galaxy's central mound is orange Antares, which gently illuminates a surrounding mistiness. Moving on some 5½ cm (2¼ inches) we find the Coalsack Nebula, a lozenge-shaped blot silhouetted against the Milky Way, with Alpha and Beta Centauri to its left. What looks like a bright star above these is actually the rich globular cluster Omega Centauri. Next comes a particularly bright patch of the Milky Way in Carina with the pinkish Eta Carinae Nebula overlain. Right of this can just be made out the large, faint loop of the Gum Nebula, an ancient supernova remnant, extending above and below the Milky Way. Canopus, the second-brightest star in the sky, lies below it. To the right of the Gum loop is the blue-white brilliance of Sirius, the brightest star of all, which heralds the stars of the constellation Orion, seen tilted on its side: Betelgeuse is the orange star 3½ cm (1.4 inches) to the right of Sirius, with Orion's belt and the Orion Nebula to its lower left. Centred in the Milky Way above Orion is the florid pink Rosette Nebula. Finally, at upper right are the stars Castor and Pollux, the celestial twins of the constellation Gemini.

The Milky Way

Piled upon each other through the effect of perspective, stars mass into dense drifts along the course of the Milky Way, dappled by patches of darker dust. The 360° panorama above, a composite of eight individual frames taken from the United States and Australia, is centred on the heart of our Galaxy, in the constellation Sagittarius. It depicts what we would see if we could float freely in space without the Earth or Sun to interrupt our view.

The Milky Way is simply the disk of our Galaxy seen from our position within. Overall, the Galaxy (given a capital G to distinguish it from other galaxies) is estimated to contain several hundred billion stars and to span more than 100,000 light years from rim to rim. On that scale, the 4-light-year distance that separates the Sun from the nearest star, Alpha Centauri, shrinks to insignificance. Arms of stars and gas coil outwards from the Galaxy's hub. Recent research suggest that the hub is not lens-shaped as was long supposed but consists of a short, straight spoke, similar to that in M83 portrayed on page 103.

With a stately grace befitting its huge girth, the Galaxy slowly turns on its axis, although not as a solid wheel. Each star follows its own orbit around the Galaxy's hub. Our Sun takes about 220 million years to complete one circuit, so it has gone around the galactic centre about 20 times since it was born. As a result of these stellar motions, imperceptible in a human lifetime, the constellations alter their shape over many thousands of years.

Our Sun, along with the bright stars visible in the night sky, lies about halfway from the Galaxy's axis to its perimeter, near the

inner edge of the same spiral arm that houses the Orion Nebula. Facing the northern constellation Cygnus, to the left of the Galaxy's central mound on the panorama above, you are looking along the local spiral arm; turn towards the southern constellations of Vela and Puppis (right of the central mound in the panorama) and you gaze along the local arm as it recedes in the opposite direction.

We can trace the nearby spiral arms from their visible content, notably star-forming nebulae, young star clusters and brilliant individual stars. To extend our picture of the Galaxy to greater distances, astronomers map the radio emission at 21-cm wavelength from the invisible hydrogen gas that lines the spiral arms. Radio waves penetrate the dark clouds of dust that obscure the long-range view at optical wavelengths. So, too, do infrared waves, which offer another method of observing the regions of our Galaxy otherwise concealed from sight. Most difficult of all to see is the galactic centre, veiled behind dense clouds of dust and gas. A radio source known as Sagittarius A* marks the dead centre. From the orbital motions of gas swirling around Sagittarius A*, astronomers conclude that it contains a black hole with a mass of a few million Suns. Astoundingly large as this may seem, the central black hole in our Galaxy is small by comparison with those found at the centres of some other galaxies.

Yet the Galaxy's greatest dark secret resides not at its heart but at its periphery. An invisible halo surrounds the Galaxy, containing ten times as much mass as the visible stars and nebulae. We know the halo is there because it reveals itself by the gravitational effect it has on the rotation of the rest of the Galaxy. Our Galaxy is far from unique: most other galaxies show signs of massive invisible halos too. What these dark halos consist of is one of the major mysteries of modern astronomy. Recent results suggest that faint, low-mass stars such as white dwarfs are a major component; black holes and subatomic particles left over from the Big Bang may also contribute to the invisible mass.

Astronomers are planning an ambitious galactic census using a star-mapping satellite called Gaia, to be launched around 2010. Gaia will chart the positions, distances and motions of a billion stars – the galactic equivalent of the Human Genome Project – to create an accurate three-dimensional map of our Galaxy that will help answer many questions regarding its structure, origin and evolution.

◀ **Cutting to the core:** At infrared wavelengths the true profile of the Galaxy is seen, cutting through the dust clouds that obscure the optical view. Most of the infrared emission mapped here by the Cosmic Background Explorer (COBE) satellite comes from stars cooler than the Sun in the Galaxy's spiral arms and central bulge.

Our Galaxy's companions

Near-spherical hives of stars known as globular clusters are dotted in a halo around our Galaxy. They are much larger than the open clusters which inhabit the Galaxy's spiral arms and also much more highly populated, containing hundreds of thousands of stars in a volume of space from a few dozen to a few hundred light years wide. Another essential difference between the two types of cluster is that open clusters are young, whereas the stars in most globular clusters are immensely old – among the oldest inhabitants of the Galaxy, in fact. Evidently, globular clusters came into being while the Galaxy was still forming, around 13 billion years ago. The antiquity of their constituent stars is revealed by their composition: almost pure hydrogen and helium, without the heavier elements produced by generations of supernova explosions that have been incorporated into younger stars such as the Sun.

About 150 globular clusters are known in our Galaxy. Rather like a scaled-up version of comets orbiting the Sun, they follow highly elongated paths around the galactic centre every 100 million years or so, diving through the galactic plane on each circuit. Gravity keeps these mighty congregations bound together, but stars leak away from time to time and some clusters may be disrupted by their repeated passages through the dense spiral arms, so there may once have been many more of them. Globular clusters are a common feature of all but the smallest galaxies. The Andromeda Galaxy, our nearby spiral neighbour, has three times as many globulars as we do, while giant elliptical galaxies are attended by thousands.

Beyond the globular clusters, our Galaxy is accompanied by a handful of mini-galaxies. The two most prominent are the irregularly shaped Magellanic Clouds, named after the Portuguese explorer Ferdinand Magellan. To the naked eye, they look like scraps torn from the Milky Way and pasted onto the celestial vault in the far-southern sky. The Large Magellanic Cloud, about 170,000 light years away, is roughly one-quarter the diameter of our own Galaxy and contains one-tenth the number of stars. The Small Magellanic Cloud, 200,000 light years distant, is in turn about one quarter the diameter and one tenth the mass of the Large Cloud.

Both Clouds orbit the Galaxy on a leisurely loop that takes billions of years. They are enveloped in a ribbon of hydrogen gas termed the Magellanic Stream, invisible to the eye but detectable by radio astronomers; this gas is thought to have been torn from the Small Cloud during a close approach between the two Clouds a few hundred million years ago. In billions of years to come, the Magellanic Clouds will probably spiral into our Galaxy and merge with it, a fate already happening to a small galaxy on the far side of the galactic centre called the Sagittarius Dwarf.

▶ **Globe light:** A million or more stars crowd into Omega Centauri, the largest and brightest globular cluster in our Galaxy, 17,000 light years from us. At its centre, stars jostle 100,000 times more closely than in the Sun's vicinity. To the naked eye this stellar throng appears as a hazy ball in the southern constellation Centaurus and was labelled as a star on early charts. It shows prominently on the Milky Way panorama on page 95, above Alpha and Beta Centauri. The cluster's elliptical outline is apparent on this photograph taken at the Anglo-Australian Observatory.

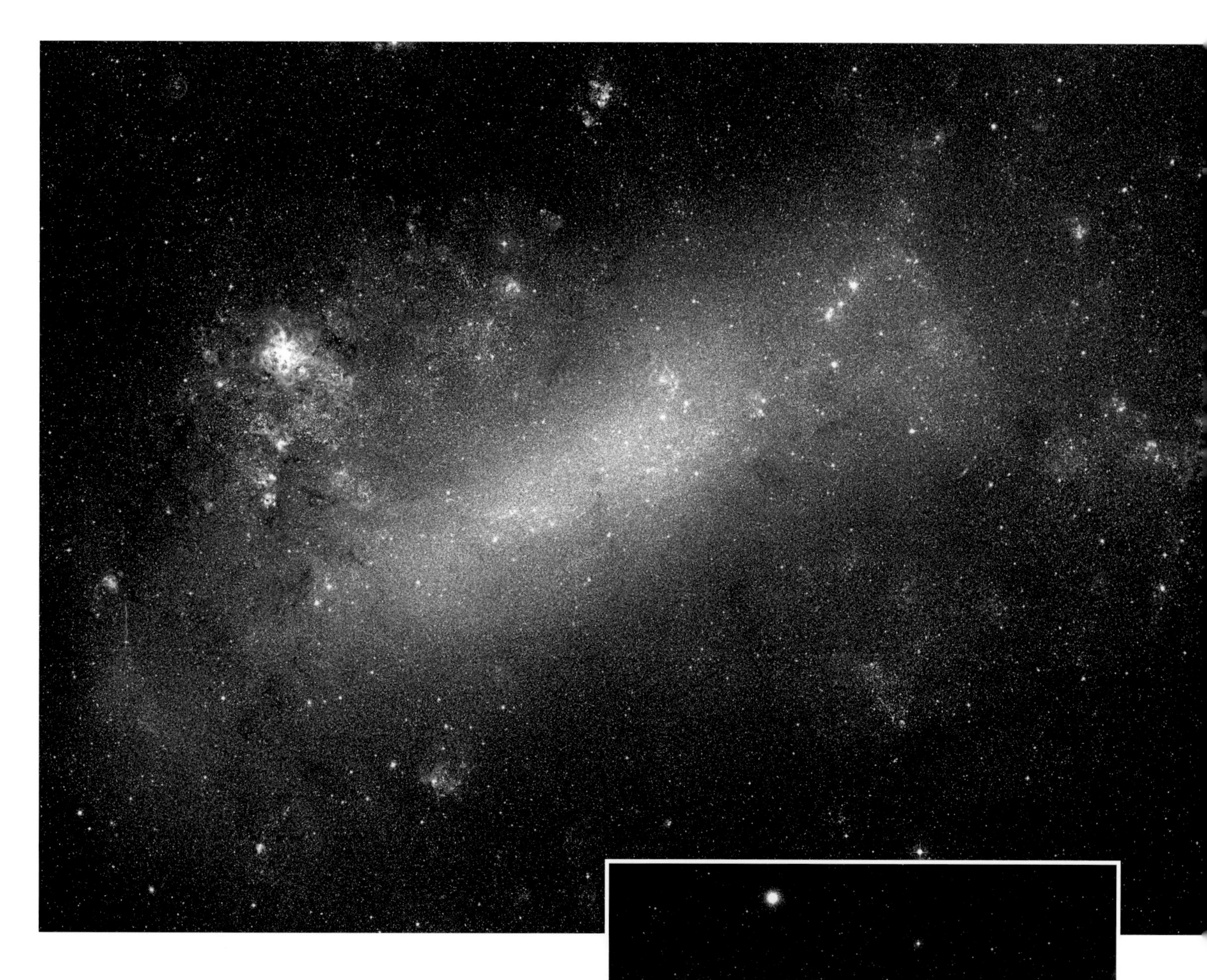

THE STARRY CLOUDS OF MAGELLAN

Seen through the eye of the United Kingdom Schmidt Telescope in Australia, an instrument that combines an ultra-wide field of view with pin-sharp clarity, the Large Magellanic Cloud, above, resolves into a swell of starlight festooned with pink gas clouds. Most prominent of the gas clouds is the Tarantula Nebula, at upper left, so named because of its spidery loops. About 1000 light years in diameter, the Tarantula is larger and brighter than any nebula in our Galaxy. Were it as close to us as the Orion Nebula, it would fill the entire constellation of Orion and cast shadows at night. The brightest supernova of modern times erupted near the Tarantula in 1987 (page 91). The Small Magellanic Cloud, right, appears somewhat scorpion-shaped, with a tail of nebulae and star clusters. The rounded "sting" at the top of the picture is NGC 362, a foreground globular cluster in our own Galaxy and not part of the Small Cloud.

The Andromeda Galaxy

Over ten times farther off than the Magellanic Clouds lies the great spiral in Andromeda, simultaneously the nearest large galaxy to us and the most distant object within reach of the human eye. Starlight currently reaching us set out from there around 2½ million years ago, a time when our ape-like ancestors roamed the plains of Africa. On clear nights, the Andromeda Galaxy is visible to the eye as a hazy smudge in northern skies for those who know where to look, between the Great Square of Pegasus and the W-shape of Cassiopeia. Binoculars enhance its faint glow, but only on long-exposure photographs can we trace the full extent of its enfolding spiral arms. It appears elongated because we view it at a steep angle, but face-on its true spiral nature would be apparent.

The Andromeda Galaxy and our own are the largest members of a modest cluster of about 40 galaxies called the Local Group, which also includes the Magellanic Clouds and Andromeda's own satellite galaxies. Although the Andromeda Galaxy is often regarded as a twin of our own, albeit somewhat larger and more massive, the two spirals exhibit subtle differences of structure and star content which hint at disparate histories. Whereas our home Galaxy is thought to have formed from the collapse of an immense gas cloud shortly after the Big Bang, supplemented by the later absorption of a few satellite galaxies, the Andromeda spiral probably grew to its present size and brightness by successive mergers with numerous smaller galaxies within the Local Group.

▶ **Giant spiral:** A coil of distant suns seen nearly edge-on, the Andromeda Galaxy is the largest member of our Local Group. Two dwarf galaxies are also visible, smaller and closer than the Magellanic Clouds that accompany our Galaxy: the rounded M32 just below centre, and the elongated M110 at upper right. This classic portrait, taken with the Schmidt Telescope at Palomar Observatory around 1960, was the first-ever in colour of the Andromeda Galaxy. In modern digital form it remains as good as any more recent image. Stars scattered across the view are foreground members of our Milky Way.

This story of galaxy growth through amalgamation is a familiar one throughout the Universe, as the photographs on later pages demonstrate. For the Andromeda Galaxy, the ultimate entanglement awaits: with our own Galaxy. Bound together by their mutual gravity, our Galaxy and the Andromeda Galaxy are approaching each other at a speed of around 500,000 km/h (300,000 mile/h), which will quicken as they close. About three billion years from now, anyone living in the Sun's neighbourhood will see their night sky filled by the Andromeda Galaxy, which will first pass by before falling back to merge with us about a billion years later. When the galaxies finally intermingle, there will be two Milky Ways arching across the sky for 100 million years or so.

Two possible fates await the Sun and Earth, depending where they lie on their circumgalactic orbit when the merger occurs. They could be flung out on a tail of stars and gas stretching into the darkness of intergalactic space. Or they could be hurled into the centre of the melding pair where where massive young stars will be forming – and dying as supernovae – in a great burst, lighting up the sky.

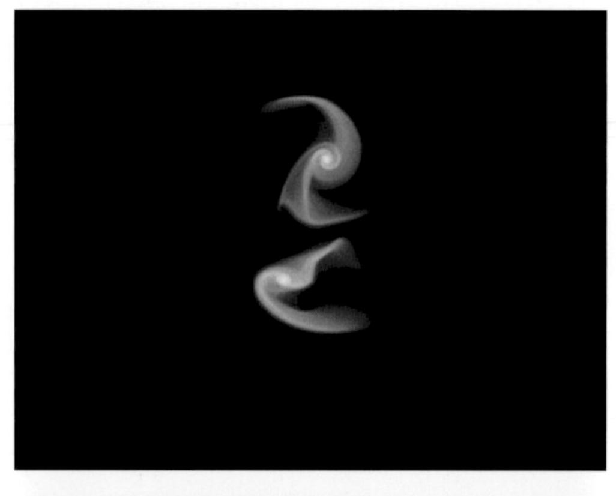

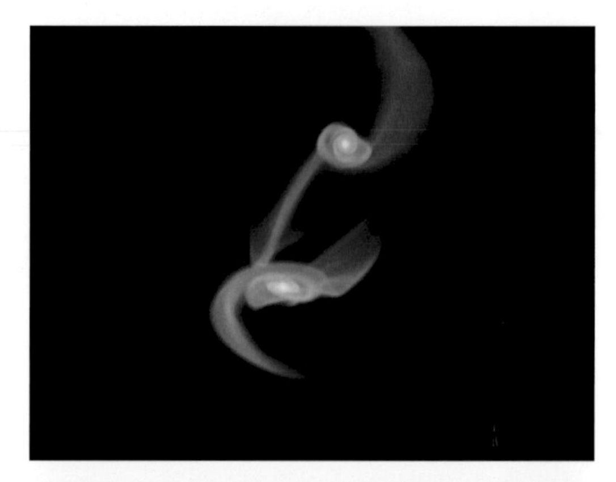

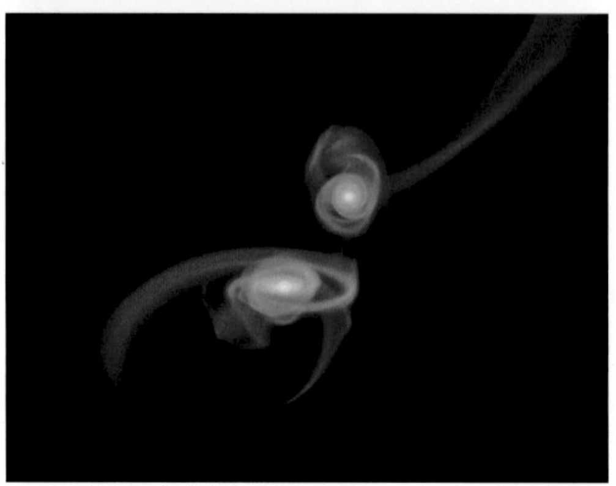

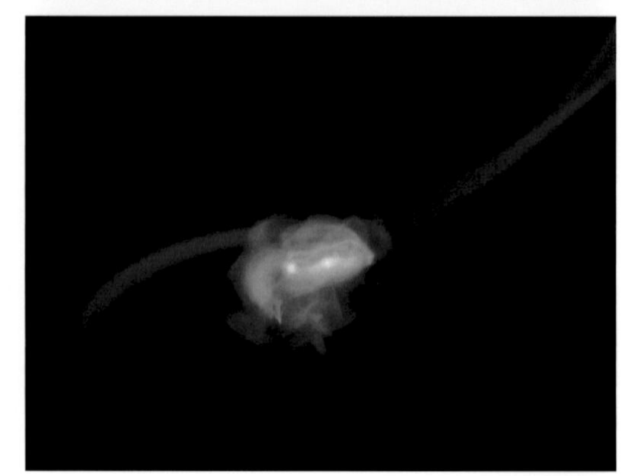

▶ **Taking the plunge:** Four stages in the fatal dance between the Andromeda Galaxy and our Galaxy that will lead to their eventual merger. These computer-generated images display a region of space 1 million light years wide, less than half their current separation.
Top left: First the partners rush past each other, flinging their arms wide. Our Galaxy (face-on, at the top) has moved upwards past Andromeda (seen at an angle).
Top right: Gravitational tides draw out a starry bridge in the widening gap between the partners. But in another few hundred million years their motion will be reversed.
Bottom left: Unable to break the gravitational leash, the partners now start to plunge back together into their final embrace.
Bottom right: Ripples of stars spread outwards as the galaxies intertwine. Eventually the coalesced couple will settle down as a giant elliptical galaxy.

Other galaxies

Magnificent galaxies of stars are dotted like islands through the sea of space, extending as far as the largest telescopes can see. They come in two main shapes: spiral, with arms, and elliptical, without arms. In addition there are irregular galaxies that do not fit either of these main categories. Many irregulars owe their distorted shapes to gravitational interplay with other galaxies.

Spiral galaxies exhibit a considerable variety of appearance. Some fling their arms wide as if in celebration while others tuck them in tightly. Spirals have masses from about a thousand million to a million million times that of the Sun, and diameters between 10,000 and over 200,000 light years. Hence our Galaxy and the Andromeda Galaxy are near the top of the range of spiral sizes.

Where spirals are tilted side-on to us we can see that their arms are compressed into a thin disk around a central mound. Stars in the central mound are usually much older than those in the arms, having formed early in the galaxy's life. By contrast, the spiral arms contain stars with a wide range of ages, because star birth has continued there throughout the galaxy's history. Even today, stars are being born from remaining pockets of gas and dust in spiral arms.

Nearly half of all spirals have a distinct inner spoke of stars and gas termed a bar, from the ends of which the spiral arms emerge. In fact, a trace of a bar can be found within most spirals. Our own Galaxy is such a case; seen from outside, it may look more like M83 shown on the facing page than the Andromeda Galaxy, which has traditionally been regarded as our near-twin. In fact, it now seems likely that bars come and go during a spiral galaxy's lifetime; rather than being oddities, they may be a natural part of spiral structure.

Although spiral galaxies are the most photogenic type of galaxy, they are not the most numerous. Bulbous ellipticals dominate the galaxy population, particularly the faint dwarf ellipticals containing only a few million stars, like oversized globular clusters. At the top of the range are the supergiant ellipticals, the largest galaxies known, stuffed with ten times as many stars as our Galaxy. Ellipticals range in shape from nearly spherical to elongated like American footballs. They are sculpted almost entirely from old stars, of similar age to those in the central bulges of spiral galaxies, and lack clouds of gas from which to make new ones.

How galaxies came into being and subsequently evolved is a major area of astronomical research. There are two rival theories: either galaxies formed from the collapse of immense clouds of gas early in the history of the Universe, or the galaxies we see today built up through successive mergers of smaller building blocks. In cases where all the available gas was turned into stars during such mergers an elliptical galaxy would have resulted; spiral galaxies arose where gas remained, or was subsequently collected, around the central bulge. Observations of distant galaxies with the Hubble Space Telescope tend to support the second picture, although both mechanisms may have been at work depending on the type of galaxy involved.

▶ **Patchy spiral:** M33 in the northern constellation Triangulum is the third-largest member of our Local Group of galaxies, about half the diameter of our own Galaxy and with one-tenth the number of stars. It has a small nucleus and broad, patchy spiral arms. This photograph was taken by the Isaac Newton Telescope on La Palma in the Canary Islands.

HOW GALAXIES GOT THEIR NUMBERS

Charles Messier, a French observer who hunted for comets in the 18th century, compiled a list of hazy-looking objects that might cause false alarms during his searches. Messier's list of comet impersonators included star clusters, gaseous nebulae and objects that we now know to be galaxies, although the existence of galaxies beyond our own was not recognized at that time. With additions by other astronomers, the compilation eventually grew to 110 entries. Objects on it are known by their list number prefixed with the letter M – for example, M1 (the Crab Nebula), M31 (the Andromeda Galaxy) and M45 (the Pleiades star cluster).

Many more star clusters and nebulous-looking objects came to light as astronomers such as William Herschel and his son John surveyed the skies with ever-larger instruments. In 1888 a Danish astronomer working in Ireland, J.L.E. Dreyer, published the *New General Catalogue of Nebulae and Clusters of Stars*, usually known simply as the NGC. With two additional *Index Catalogues* that followed, Dreyer listed over 13,000 objects; these are now known by their NGC or IC designations. Only in the 1920s, when Edwin Hubble established that certain so-called nebulae were in fact composed of distant stars, did it become clear that the majority of objects in the NGC and its supplements were actually galaxies.

▲ **Swirling:** A vortex of stars, gas and dusty sediments, M83 is an impressive spiral galaxy whose two main arms straighten at the centre to create a bar. Unusually, a third arm also emerges from the nucleus. Like all spiral galaxies, M83's bulbous centre consists of older stars which imbue it with a creamy tone, while the arms are highlighted by younger blue stars and mallow-pink clouds of hydrogen gas. Our own Galaxy, which has a modest central bar, might look like this if seen from above one of its poles. M83 lies about 20 million light years away in the southern constellation Hydra and is shown here in a true-colour portrait taken by the Anglo–Australian Telescope.

▶ **Giant elliptical:** M87 is a giant elliptical galaxy at the heart of the Virgo Cluster, 55 million light years away. Outwardly it appears as a smooth and featureless ball of old stars, resembling the central bulge of a spiral galaxy without any arms. M87 contains more stars than our entire Galaxy, and is surrounded by a swarm of thousands of globular clusters, visible as hazy spots on this photograph taken with the Anglo–Australian Telescope. Two small companion galaxies are also visible to its lower right.

◀ **Cosmic cobweb:** In an extraordinary superposition, a face-on spiral galaxy, NGC 3314A, lies exactly in front of a larger one, NGC 3314B, which is tilted at an angle. Dust in the foreground galaxy's spiral arms is silhouetted like a dark cobweb against the brighter background, while the outer spiral arms appear bright where they are projected against the blackness of space. An orange area near the centre of the image is the partially obscured nucleus of the background galaxy, reddened by intervening dust. This pair of galaxies lies about 140 million light years away in the constellation Hydra.

▶ **Drum roll:** M101 is a spiral galaxy larger than our own, with arms that unreel loosely from a small central drum. There is a noticeable asymmetry in the spread of its arms, seemingly due to the gravitational tug of a small companion galaxy outside the field of view of this portrait taken by the Isaac Newton Telescope on La Palma in the Canary Islands. M101 lies about 20 million light years away in Ursa Major.

▲ **Warp factor:** Many spiral galaxies show a slight warping of their arms due to the gravitational influence of companions, but in this galaxy, ESO 510–13, the warp is exceptional. An encounter with a smaller galaxy that is now being swallowed up has twisted the dusty arms so that their profile resembles a wavy hat brim, as seen by the Hubble Space Telescope. Over millions of years the warp will settle down into a flat plane like in the famous Sombrero galaxy (see over page). This remarkable galaxy lies about 150 million light years away in Hydra.

▶ **Long view:** Even at a distance of more than 60 million light years, the Hubble Space Telescope can detect the brightest individual stars in galaxies such as NGC 4414 in Coma Berenices, a handsome spiral whose arms are mottled by dark clouds of interstellar dust.

▲ **Hat trick:** This distinctive galaxy is popularly known as the Sombrero from its resemblance to a wide-brimmed Mexican hat. It is in fact a spiral with tightly huddled arms enfolding an unusually large central bulge. The Sombrero, also known as M104, may result from one or more galactic mergers. It lies about 40 million light years away in Virgo. This portrait was taken by the Anglo-Australian Telescope.

▶ **Barred:** Resembling a rotating propeller, NGC 1300 is a classic example of a barred spiral galaxy, with two starry arms curving outwards from the ends of a central spoke. NGC 1300 lies in the southern constellation Eridanus and is pictured here by the Anglo-Australian Telescope.

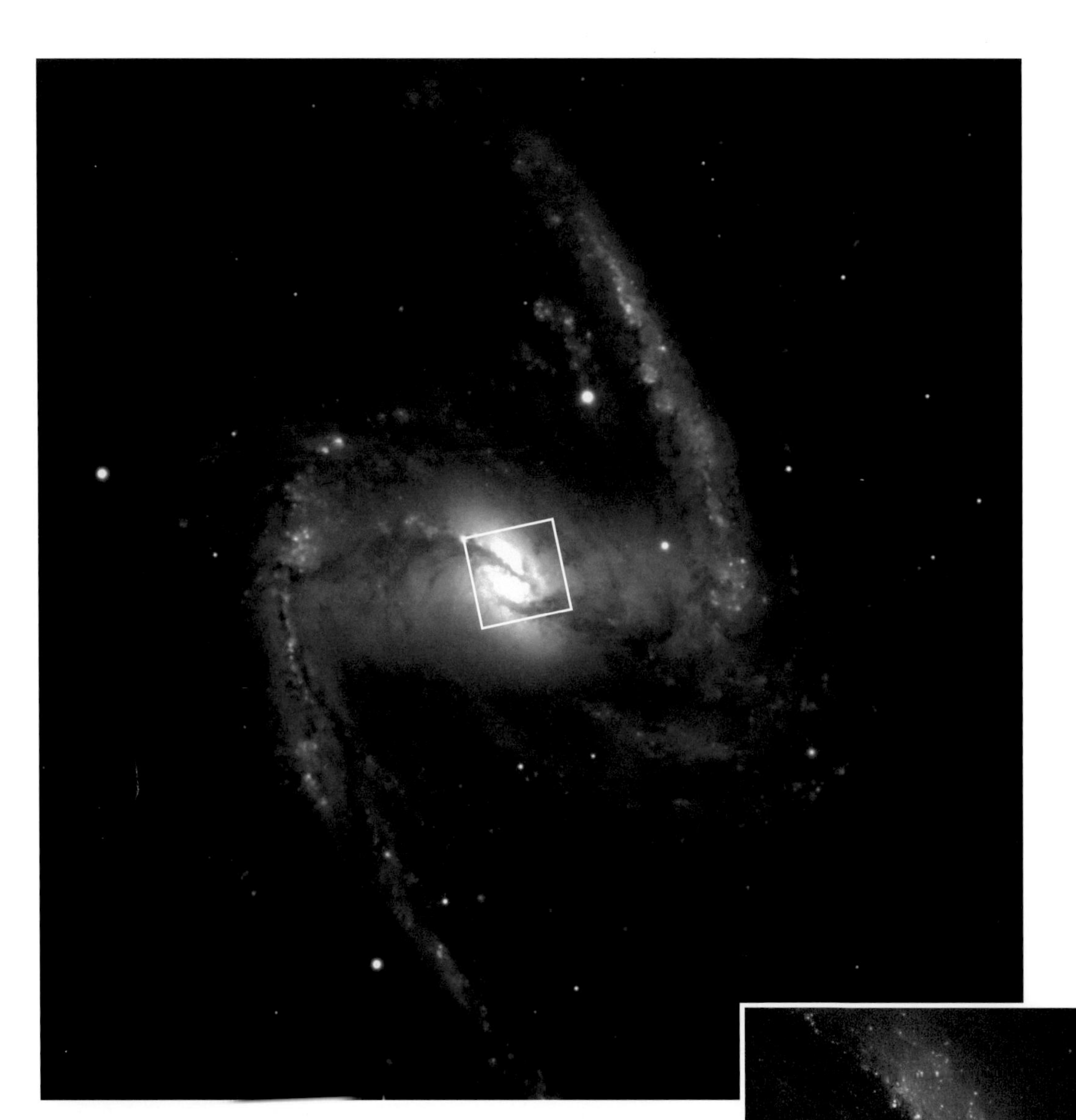

◀ **Going with the flow:** Gas flows along bars into the heart of barred spirals, where it makes new stars and enlarges the galaxy's central bulge. Only with the superior vision of the Hubble Space Telescope have astronomers been able to study this process in detail. At left, the prominent barred spiral NGC 1365, some 65 million light years away in the southern constellation Fornax, is portrayed in true colour by the Very Large Telescope in Chile. Below, the Hubble Space Telescope zooms in on the turbulent central area outlined by the white box, where bright young star clusters glitter between darker veins of dust. In the small image at bottom right, Hubble's infrared camera penetrates the dust around the burnished core to reveal more clusters of young stars, shown in red, which are obscured from view in the visible-light image. The core itself probably contains an ultra-massive black hole which is being fed by the inflowing gas.

◀ **Ring of stars:** Around the core of the barred spiral galaxy NGC 1512 the Hubble Space Telescope spies a ring of star clusters newly born from gas streaming inwards along the bar. Some clusters, obscured by dust, appear reddish.

CENTRAL BLACK HOLES

Most large galaxies have gaping black holes at their centres with masses between 100,000 and several billion times that of the Sun. Enormous as such masses may seem, they are still about 500 times less than the mass of the central bulge of each galaxy, be it spiral or elliptical. The only galaxies without a detectable black hole at their centre are those that lack an appreciable bulge, such as the spiral M33 in the Local Group.

Although the black holes themselves cannot be seen, we can detect their influence: a massive black hole's immense gravitational pull accelerates stars and gas in its vicinity to high speed. The faster the speed at which matter orbits a galaxy's core, the greater the mass of the unseen hole embedded within. Such velocity measurements have revealed that our own Galaxy has a central black hole with a mass of some three million Suns, while the mass of the black hole at the heart of the giant elliptical galaxy M87 is a thousand times greater. These black holes were probably not born this big but grew by the influx of gas and stars early in the galaxy's life, as well as through mergers with other galaxies that contained their own central black holes. Disks of hot gas around massive black holes explain the ultra-bright centres of certain galaxies and the exceptionally luminous objects known as quasars (see page 118).

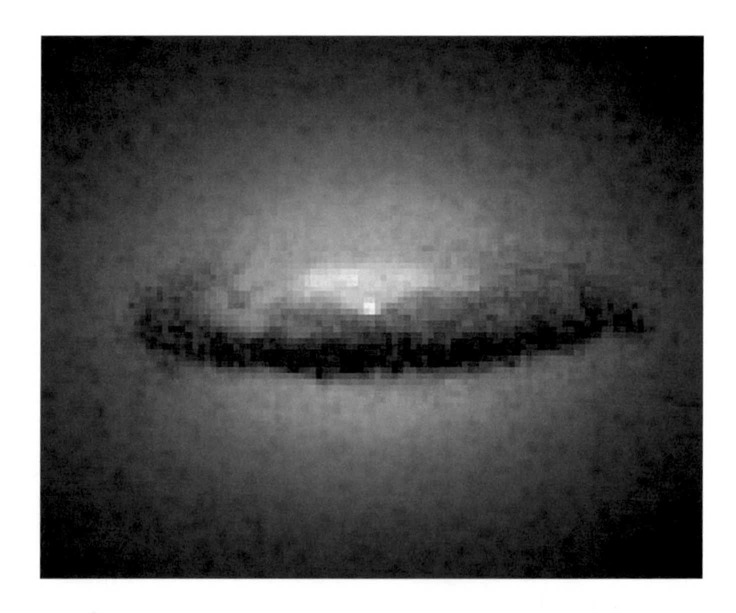

▲ **Hollow centre:** A ring of gas and dust 3700 light years in diameter orbits a black hole at the centre of the elliptical galaxy NGC 7052. From measurements of the speed of the gas, astronomers estimate the black hole's mass to be 300 million times that of our Sun. The bright spot at the very centre of the disk is the combined light of stars and hot gas crowded around the black hole. NGC 7052 lies 190 million light years away in the constellation Vulpecula.

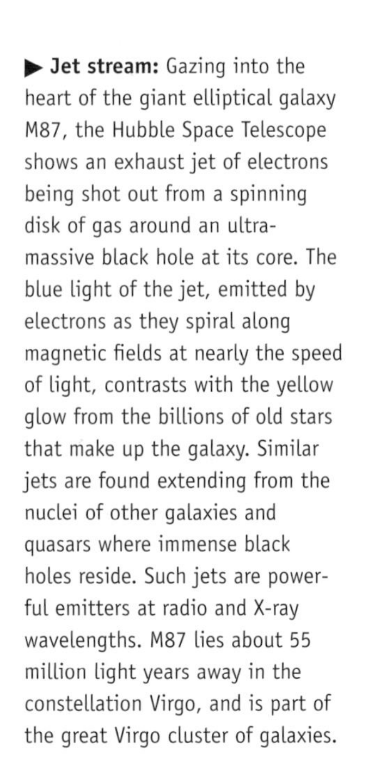

▶ **Jet stream:** Gazing into the heart of the giant elliptical galaxy M87, the Hubble Space Telescope shows an exhaust jet of electrons being shot out from a spinning disk of gas around an ultra-massive black hole at its core. The blue light of the jet, emitted by electrons as they spiral along magnetic fields at nearly the speed of light, contrasts with the yellow glow from the billions of old stars that make up the galaxy. Similar jets are found extending from the nuclei of other galaxies and quasars where immense black holes reside. Such jets are powerful emitters at radio and X-ray wavelengths. M87 lies about 55 million light years away in the constellation Virgo, and is part of the great Virgo cluster of galaxies.

▼ **Group portrait:** Hickson Compact Group 87 in Capricornus contains four galaxies of varied shape. The largest, at bottom, is an edge-on spiral galaxy, interacting with an elliptical galaxy to the right. Two smaller spirals are seen at different angles in this Hubble Space Telescope view. Bright stars are foreground objects in our own Galaxy.

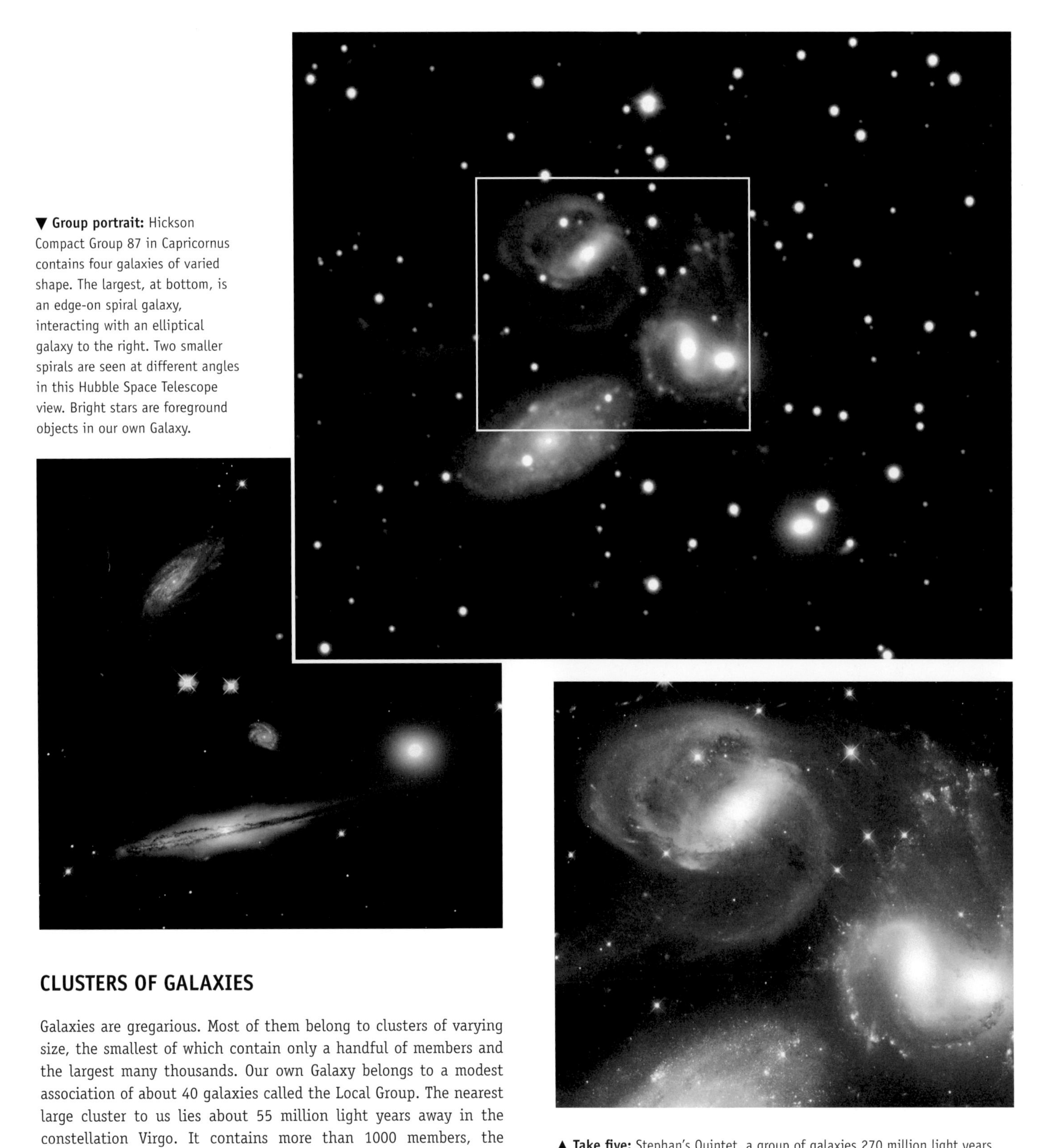

CLUSTERS OF GALAXIES

Galaxies are gregarious. Most of them belong to clusters of varying size, the smallest of which contain only a handful of members and the largest many thousands. Our own Galaxy belongs to a modest association of about 40 galaxies called the Local Group. The nearest large cluster to us lies about 55 million light years away in the constellation Virgo. It contains more than 1000 members, the largest of which is the giant elliptical galaxy M87. Such giant ellipticals are common at the centres of large clusters, whereas spirals are rare in them. This is thought to be because the spirals in such clusters have merged to form the giant ellipticals.

▲ **Take five:** Stephan's Quintet, a group of galaxies 270 million light years away in the northern constellation Pegasus, seen from Kitt Peak Observatory (top), with a close-up of the boxed section from the Hubble Space Telescope above. The leftmost of the two galaxies at lower right is barrelling its way past the other; brilliant blue star clusters are forming where the two collide.

Measuring the Universe

Unreachable though stars and galaxies may be, we can still find how far away they are. Triangulation, the traditional method of surveyors, is the first step. Astronomers take bearings on a star six months apart, when the Earth is on opposite sides of the Sun, using the diameter of the Earth's orbit as the baseline. The change in the star's bearing when seen from each end of the baseline, termed parallax, reveals its distance, the difference being greatest for the closest stars. Even so, the differences are so minuscule that they are difficult to measure reliably from the ground, due to the unsteadiness of the atmosphere through which we must look. In the 1990s a position-measuring satellite called Hipparcos, taking advantage of the pin-sharp view from space, increased the range and precision of our parallax measurements and a future satellite, Gaia, will extend that stellar harvest as far as the galactic centre.

To establish distances to other galaxies, whose stars are beyond the range of parallax, astronomers use a variety of rangefinders. Most notable among these are stars known as Cepheids which rise and fall in brightness in regular cycles lasting a few days or weeks. What makes Cepheids so useful is that the brighter they are, the longer they take to vary. Hence, by timing a Cepheid's cycle of variation astronomers can deduce its inherent luminosity, which is like the wattage of a light bulb. Even the most brilliant light bulb will appear faint if it is far away – so, once calibrated, the apparent brightness of Cepheids can be used to estimate the distance of the host galaxy. One of the prime tasks of the Hubble Space Telescope was to detect Cepheid variables in galaxies in the Virgo cluster, much farther than is possible from Earth, to establish an improved distance scale for the Universe. An example is shown opposite.

Beyond the range at which Cepheids can be seen astronomers turn to other standards, such as the brightness of supernova explosions and comparisons of the size and brightness of individual galaxies with those closer to home. With distances firmly established in these various ways, astronomers can make reliable estimates of the rate of expansion of the Universe, a figure known as Hubble's constant. According to measurements made with the Hubble Space Telescope, the expansion of the Universe carries galaxies away from us at a speed which increases by about 70 km/s (43 mile/s) with every million parsecs (a parsec being the distance at which an object would show a parallax of one second of arc – roughly 3.2616 light years). In turn, this figure tells us roughly how long must have elapsed since the Big Bang occurred that initiated the expansion. The answer is about 13 billion years.

▼ **Milestone:** Over 100 million light years away, NGC 4603 in the Centaurus cluster of galaxies is the most distant galaxy in which Cepheid variable stars have been detected. Even with the Hubble Space Telescope, special processing was needed to identify the Cepheids at such a vast distance. Bright stars in this Hubble image are foreground objects in our own Galaxy.

▶ **Fantastic light:** The penetrating glare of a supernova (at lower left of this image taken by the Hubble Space Telescope) beams from the outskirts of the edge-on spiral galaxy NGC 4526, rivalling the brilliance of the galaxy's entire nucleus. Astronomers use the exceptional luminosity of supernovae to estimate the distances to remote galaxies where normal stars cannot be seen. NGC 4526 lies in the Virgo cluster, the nearest large cluster of galaxies to us, which serves as one of the stepping stones in our distance scale of the Universe.

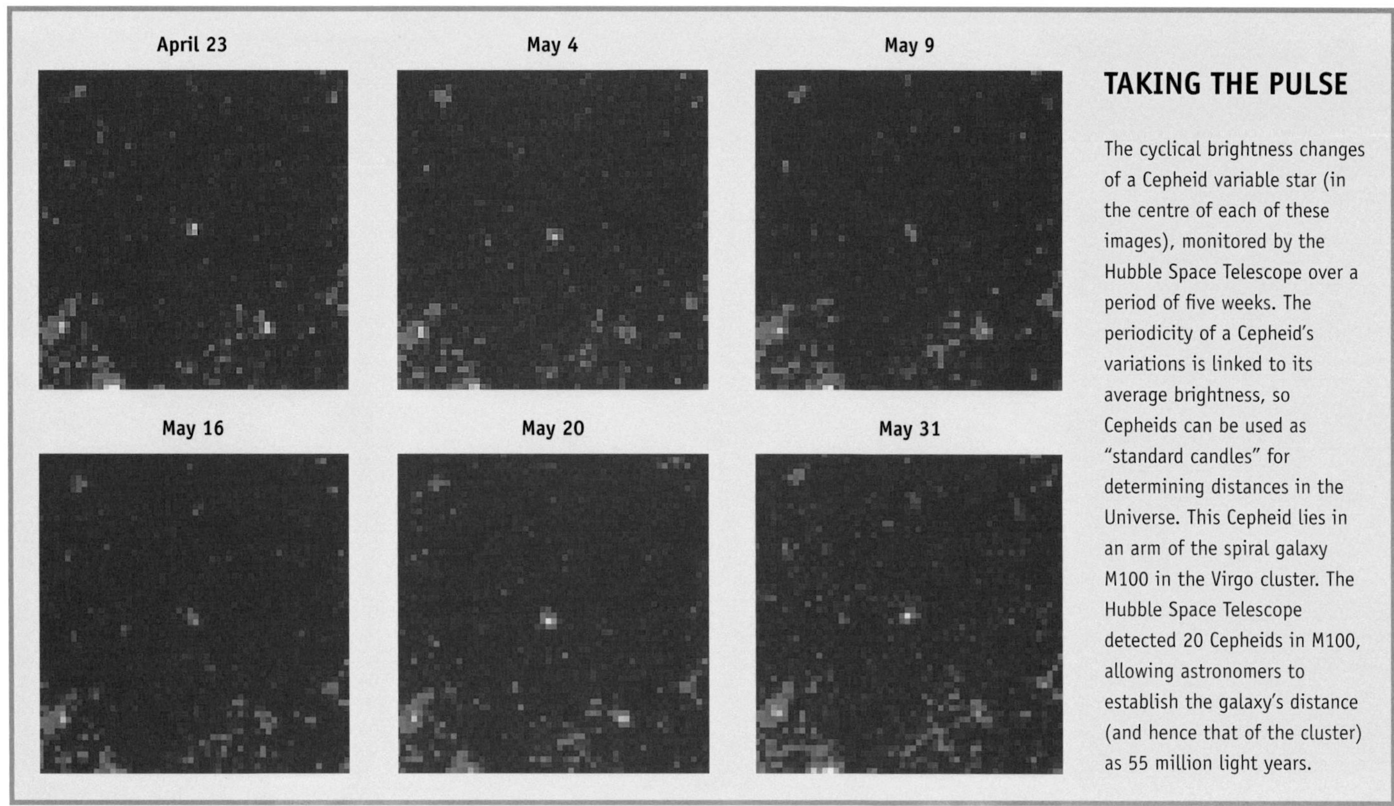

TAKING THE PULSE

The cyclical brightness changes of a Cepheid variable star (in the centre of each of these images), monitored by the Hubble Space Telescope over a period of five weeks. The periodicity of a Cepheid's variations is linked to its average brightness, so Cepheids can be used as "standard candles" for determining distances in the Universe. This Cepheid lies in an arm of the spiral galaxy M100 in the Virgo cluster. The Hubble Space Telescope detected 20 Cepheids in M100, allowing astronomers to establish the galaxy's distance (and hence that of the cluster) as 55 million light years.

Galactic collisions and mergers

Galactic encounters are the most spectacular traffic accidents in the Universe. A collision between two galaxies implies catastrophe on an unimaginable scale, but no real destruction is involved since stars are so widely spaced that galaxies can pass through each other like phantoms. Although stars do not run into each other, gas clouds most certainly will, triggering bursts of brilliant star formation. Near-misses in which the two galaxies brush past each other are more common than direct collisions, but eventually lead to mergers. During such interactions, gravity distorts the shape of the galaxies, often pulling out fantastic festoons of stars and gas. Passing galaxies are slowed by these glancing encounters and subsequently fall back together under the attraction of their mutual gravitational pulls, sealing their merger.

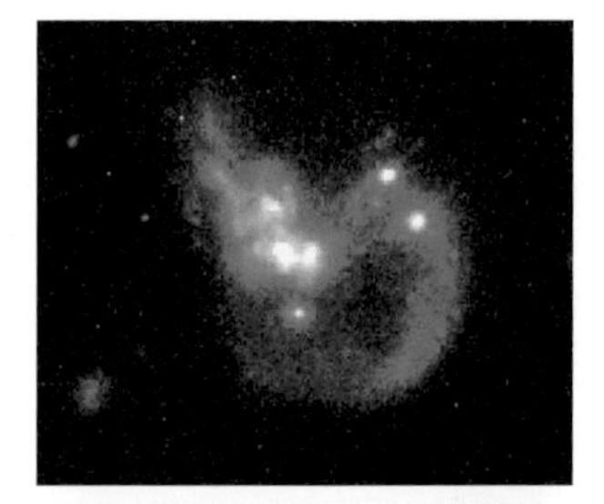

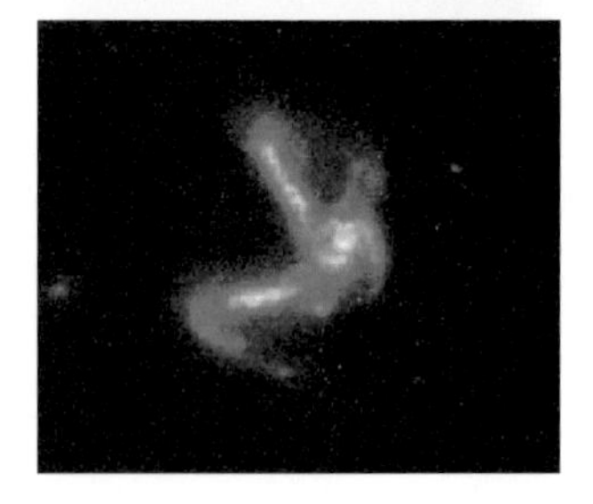

▶ **Piling in:** Multi-galaxy pile-ups billions of light years away have been spied by the Hubble Space Telescope. The mêlée at upper right appears to contain the nuclei of several galaxies. Below right, at least three distorted galaxies are converging on one spot. These views offer an insight into conditions in the early Universe, when galaxy collisions were more commonplace than they are today.

For many years, astronomers assumed that galaxies lived largely independent and untroubled lives; now, it is recognized that interactions and mergers are not only common but have played a major role in the development of galaxies and the formation of stars within them. Most large galaxies may experience several interactions during their lifetimes which can completely change their appearance – for example, the merger of two similar-sized spirals can produce an elliptical galaxy, as is believed will happen when our own Galaxy and the Andromeda Galaxy combine in billions of years' time. However, an interaction or merger takes hundreds of millions of years to complete, so in each case we see only snapshots of the activity. The following pages present a gallery of galaxies either undergoing interactions or showing the results of past encounters. Astronomers are still struggling to understand the processes involved in some of the most extreme examples.

▼ **Coupling:** Gravitational forces from the spiral galaxy NGC 2207 (left) have disrupted the smaller IC 2163, pulling out a streamer of stars and gas that stretches for 100,000 light years to the right of this Hubble Space Telescope image. IC 2163 is swinging counterclockwise past NGC 2207 and currently lies behind its spiral arms. Trapped in mutual embrace, these two galaxies will continue to distort and disrupt each other before eventually merging.

CANNIBAL GALAXY

Centaurus A, also known as NGC 5128, is a giant elliptical galaxy embellished with a dark mane of dust, apparently the remains of a smaller spiral it has swallowed in an act of galactic cannibalism. Fuelled by gas from the devoured galaxy, an ultramassive black hole at the centre of Centaurus A is ejecting atomic particles in high-speed jets. These are invisible in photographs like that above, taken with the Anglo-Australian Telescope, but they emit powerfully at radio and X-ray wavelengths. At right, the Hubble Space Telescope zooms in on the galaxy's glowing nucleus, veiled and reddened behind the dust lane which is frosted with blue new-born stars. At a distance of 11 million light years, Centaurus A offers astronomers their closest look at the outcome of a major galactic merger.

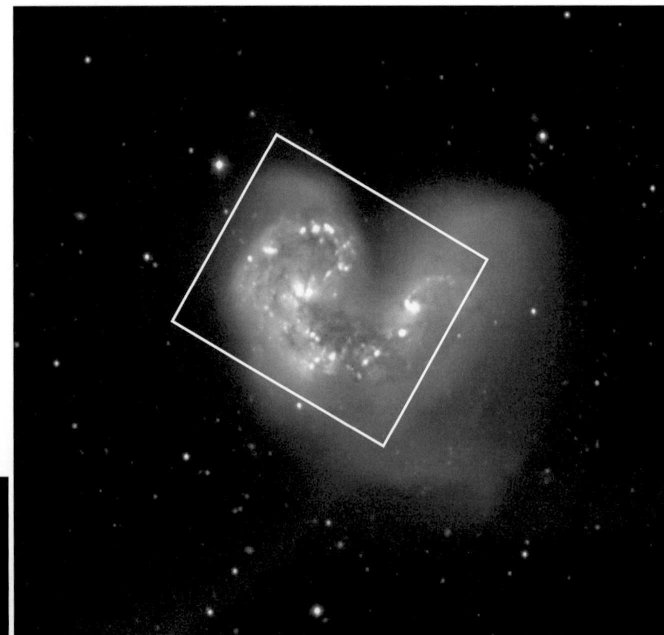

THE ANTENNAE

The archetypal pair of colliding galaxies is NGC 4038 and 4039, popularly known as the Antennae because of long streamers of stars and gas torn off from them resemble an insect's feelers. At left, the two distorted spirals, 63 million light years away in the constellation Corvus, are shown by the Anglo-Australian Telescope. The central area in the white box is depicted in greater detail below by the Hubble Space Telescope. On Hubble's pictures, astronomers have identified over a thousand bright blue star clusters bursting into life where immense clouds of gas have been squeezed by the collision. The largest of the new clusters contain a million stars, similar to the size of globular clusters in our own Galaxy.

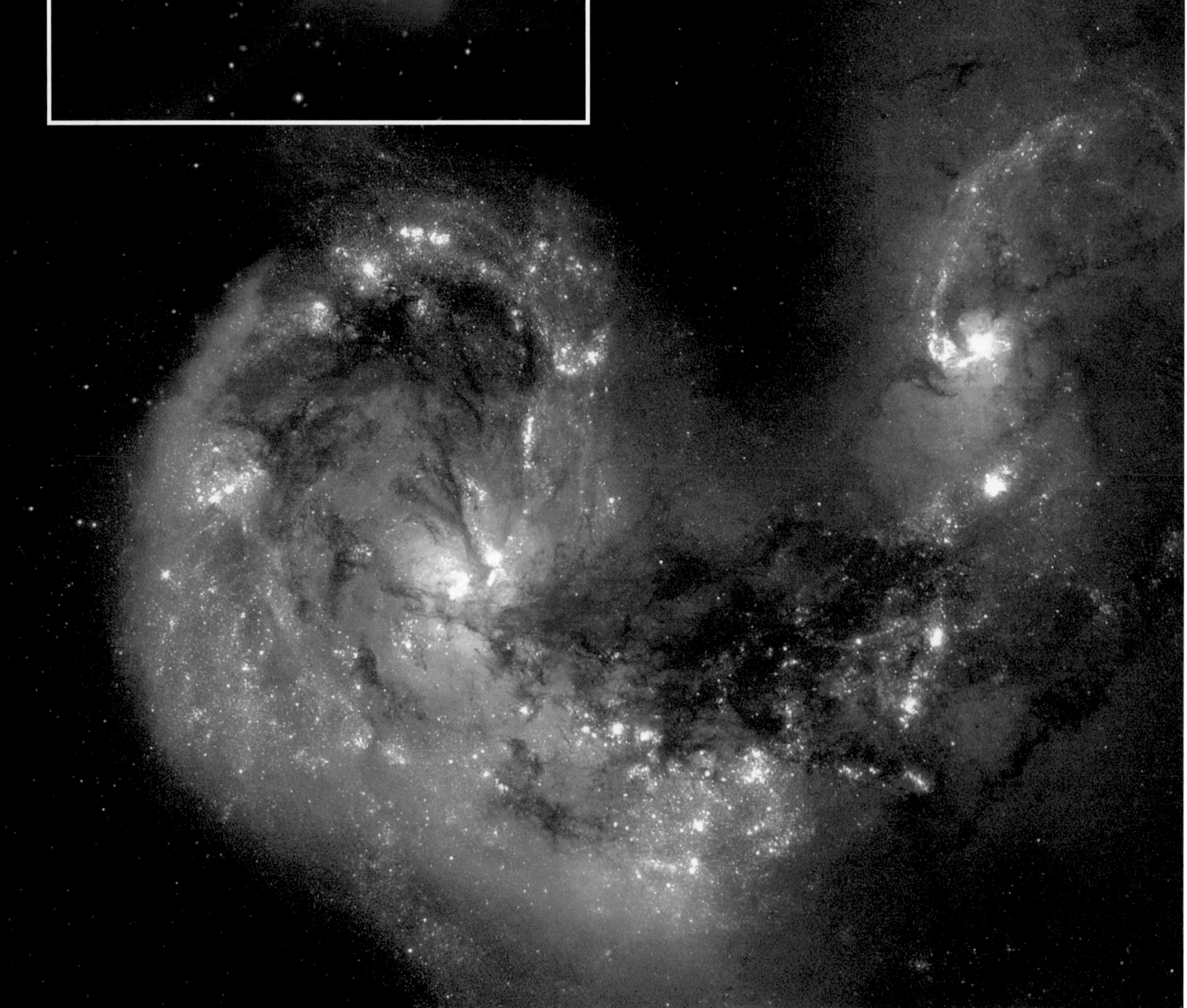

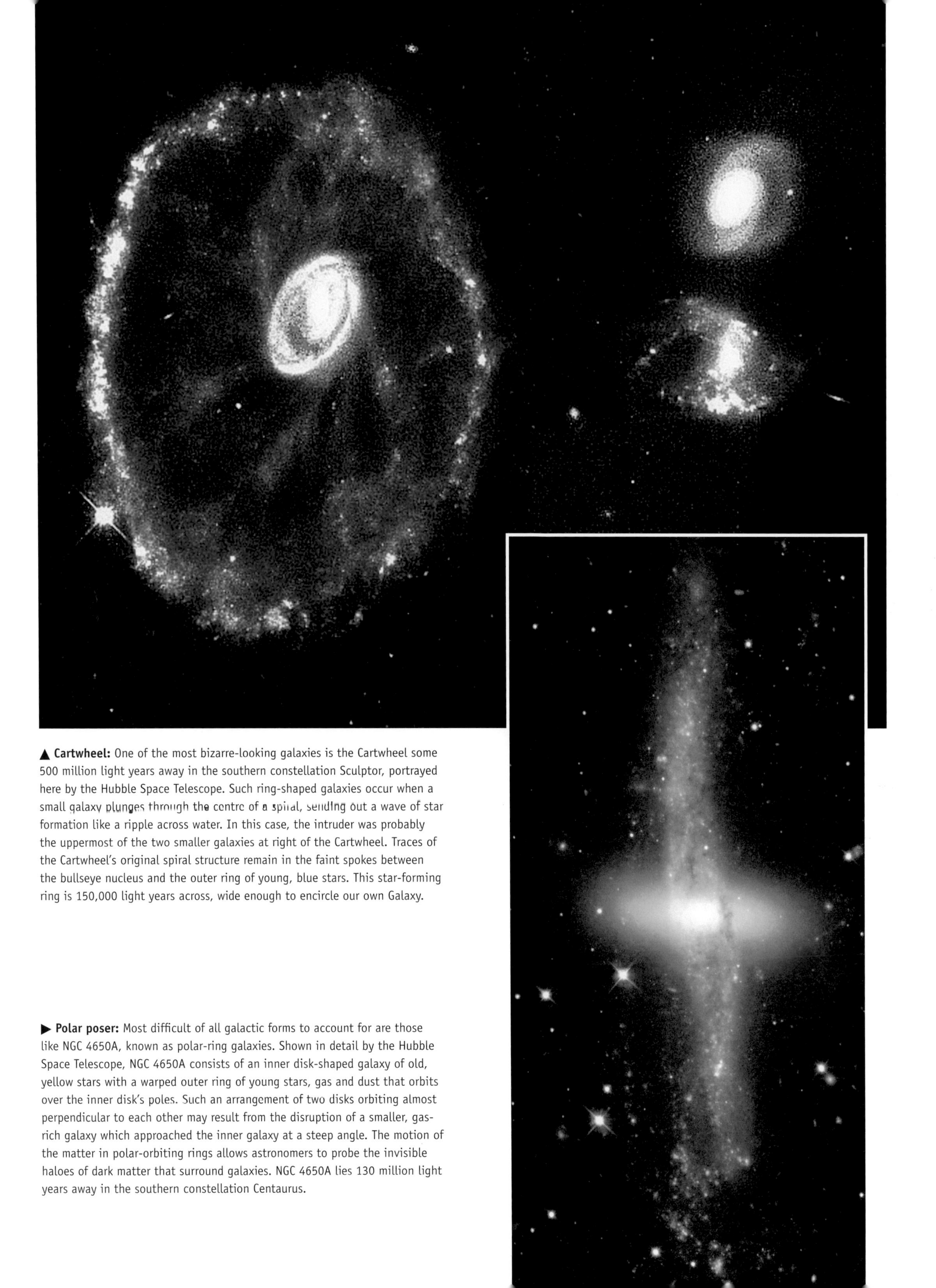

▲ **Cartwheel:** One of the most bizarre-looking galaxies is the Cartwheel some 500 million light years away in the southern constellation Sculptor, portrayed here by the Hubble Space Telescope. Such ring-shaped galaxies occur when a small galaxy plunges through the centre of a spiral, sending out a wave of star formation like a ripple across water. In this case, the intruder was probably the uppermost of the two smaller galaxies at right of the Cartwheel. Traces of the Cartwheel's original spiral structure remain in the faint spokes between the bullseye nucleus and the outer ring of young, blue stars. This star-forming ring is 150,000 light years across, wide enough to encircle our own Galaxy.

▶ **Polar poser:** Most difficult of all galactic forms to account for are those like NGC 4650A, known as polar-ring galaxies. Shown in detail by the Hubble Space Telescope, NGC 4650A consists of an inner disk-shaped galaxy of old, yellow stars with a warped outer ring of young stars, gas and dust that orbits over the inner disk's poles. Such an arrangement of two disks orbiting almost perpendicular to each other may result from the disruption of a smaller, gas-rich galaxy which approached the inner galaxy at a steep angle. The motion of the matter in polar-orbiting rings allows astronomers to probe the invisible haloes of dark matter that surround galaxies. NGC 4650A lies 130 million light years away in the southern constellation Centaurus.

◀ **Starburst:** M82, an edge-on spiral galaxy, is undergoing an immense surge of star formation as a result of an encounter about 600 million years ago with a larger neighbour, M81, which is out of this picture. The starburst may be fuelled by debris from M82 itself that has slowly fallen back on the galaxy since the interaction. M82's centre is bursting with hot, young stars and supernova remnants. Radiation from these energetic sources is driving sprays of hydrogen gas outwards for more than 10,000 light years from the galaxy. In this false-colour image from the Japanese Subaru Telescope on Mauna Kea, Hawaii, the hydrogen gas is shown red. M82 lies about 12 million light years away in Ursa Major. The white streak at lower right is due to an overexposed star image.

▶ **Mighty Mice:** Luminous tails of stars and gas stretch 100,000 light years into intergalactic space from two intertwined spirals nicknamed the Mice, seen here by the Hubble Space Telescope's Advanced Camera for Surveys. Computer reconstructions conclude that the tails were wrenched out by gravitational interactions when the galaxies swerved past each other about 160 million years ago. The tail at right is long enough to reach from one side of our own Galaxy to the other. Clumps of stars and gas within it may eventually become free-floating mini-galaxies. The interacting pair, also known by the catalogue number NGC 4676, will merge around 400 million years from now. They are part of a cluster of galaxies some 300 million light years from us in the northern constellation Coma Berenices.

Foaming with stars, the Whirlpool Galaxy (also known as M51) is a magnificent spiral 28 million light years away in the constellation Canes Venatici, portrayed here by the William Herschel Telescope on La Palma in the Canary Islands. A small satellite galaxy has brushed past the Whirlpool in the past few hundred million years, pulling out and twisting one of its arms. The companion now lies behind the arm.

Quasars

Among the Hubble Space Telescope's most significant achievements has been to clarify the nature of quasars, remote powerhouses which baffled astronomers for a generation following their discovery in the early 1960s. The puzzle was this: quasars appeared starlike (their name is an abbreviation of "quasi-stellar") yet their speed of recession, if due to the expansion of the Universe, placed them farther off than any galaxy then known. To be visible at such distances they would have to be more luminous than a normal galaxy. Yet they varied in brightness in a matter of weeks or months which meant that they must be small enough for light to cross them in that time – in other words, their diameters were only a few light weeks or months, considerably less than the four light years that separates our Sun from Alpha Centauri. This perplexing combination of overwhelming power and diminutive size was the essence of the quasar paradox. It led some astronomers to speculate that quasars were not as far away as their velocities indicated, or even that some entirely new physical processes might be at work.

▶ **Homing in on quasars:** The home galaxies of four quasars are revealed here by the Hubble Space Telescope. From the top, quasar PG 0052+251 is found to reside at the centre of a spiral galaxy, whereas PHL 909 lies within an elliptical galaxy. PG 1012+008 is merging with the bright galaxy below it; the compact galaxy at left of the quasar may also be about to join in the merger. The bottom view is of PKS 2349, which is swallowing a small galaxy similar in size to the Large Magellanic Cloud, just visible directly above it. The four quasars and their host galaxies are about 1.5 billion light years from us.

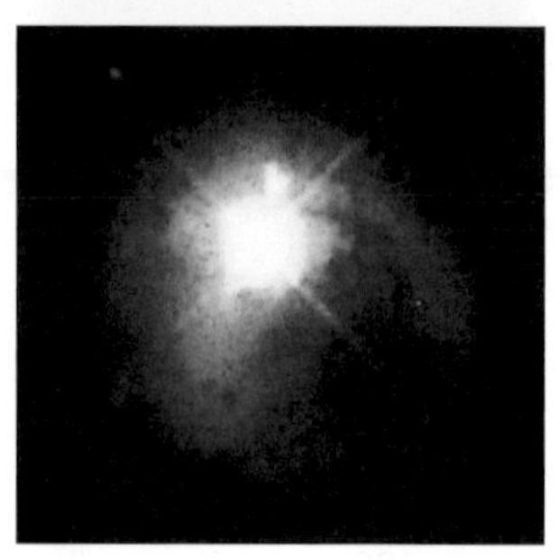

◀ **Brightest quasar:** 3C 273, some three billion light years away in Virgo, is the brightest quasar as seen from Earth. Shooting out from its core is a high-speed jet of atomic particles, shown here in a false-colour image from the Hubble Space Telescope. This jet seems to be identical in nature to the one emerging from the core of the giant elliptical galaxy M87 (page 108), demonstrating a generic link between quasars and galaxies with hyperactive centres.

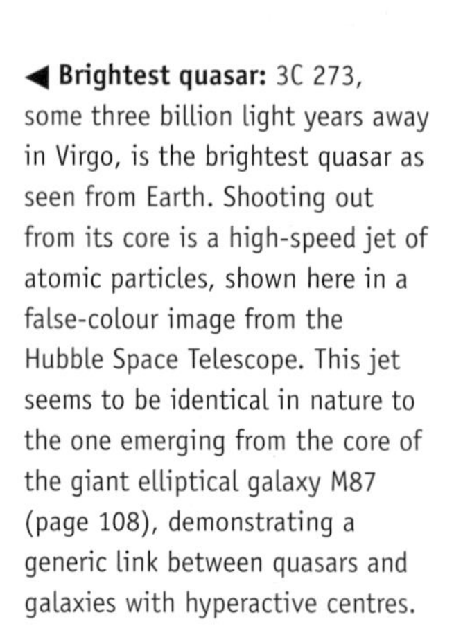

SEYFERT GALAXIES – RELATIVES OF QUASARS

Seyferts are galaxies with a pronounced twinkle in their eye, like scaled-down versions of quasars. They are named after the American astronomer Carl Seyfert who identified the first examples in 1943, long before their true nature was understood. Seyferts are more numerous than quasars but less powerful. At right, the Hubble Space Telescope shows an area about 1000 light years wide at the heart of a Seyfert galaxy some 13 million light years away in the southern constellation Circinus. A cone of gas, shown whitish-pink in this false-colour image, is being ejected by a massive but invisible black hole at the galaxy's brilliant core. A similar cone is probably being ejected in the opposite direction but is obscured from view by intervening gas and dust in the galaxy's plane. Gas expelled previously from the black hole's environs forms magenta-coloured streamers at the top of the image. Like the majority of Seyferts, the Circinus galaxy is a spiral.

Telescopes on Earth were unable to see quasars clearly enough to solve the conundrum. But the Hubble telescope has confirmed the widely held belief that quasars lie at the heart of distant galaxies and, furthermore, that they are related to other types of galaxy with hyperactive centres such as Seyferts (see box above) and radio galaxies. In all such cases the seat of power is a central black hole with a mass of millions or billions of Suns. The black hole sits like the eye of a hurricane within a circling disk of dust and gas garnered from passing nebulae and stars that have been shredded by the black hole's mighty gravitational grip. Gas falling towards the black hole heats up to temperatures hundreds of times that of the surface of the Sun, accounting for the small but brilliant cores of these objects. The luminosity of the nucleus depends on the mass of the central black hole and the amount of hot gas around it. Quasars are highly luminous because they are being fed by gas from a galactic merger, as the Hubble photographs have shown. Quasars were common early in the history of the Universe because interactions were then a regular feature of galactic existence.

Not all of the gas is swallowed by the black hole. Some of it is ejected at velocities approaching that of light along or close to the rotation axis of the disk, creating jets and radio-emitting lobes that can stretch for millions of light years either side of the galaxy. In cases where the disk is oriented nearly edge-on to us, the central brilliance is obscured at optical wavelengths but the intense radio emission from the core and surrounding lobes can still be detected by radio telescopes. Such an object is termed a radio galaxy; a famous example is Centaurus A, shown on page 113.

As the black hole gradually consumes the surrounding gas, the quasar's brightness declines, although it can be boosted again by the injection of fresh fuel through a new merger. Many galaxies, quiet today, harbour a central black hole now starved of fuel – the dormant core of a quasar that once blazed brilliantly. Rather than being freaks as was originally thought, quasars are now recognized as the most extreme examples of the activity experienced by many galaxies at some time in their lives.

Gravitational lensing

Albert Einstein revolutionized our understanding of gravity in 1915 when he described it as a curvature of space, not an attraction between masses as it had been thought of since the time of Isaac Newton. Curved space bends the paths of all objects passing through it – planets, for example, are kept in their orbits by the curvature of space around the Sun. What's more, Einstein's theory predicted a completely unanticipated effect – the path of light rays is bent by curved space, too. When light dips in and out of an exceptionally strong gravitational field, such as around a black hole or within a group of galaxies, the extreme curvature of space can act upon it like a distorting lens, smearing it into surreal arcs or cleaving it into two or more separate images.

The most prominent example of such a gravitational lens was discovered in 1985 by Earth-based telescopes. Called the Einstein Cross, it consists of four magnified images of a distant quasar refracted by the nucleus of a spiral galaxy that by chance lies directly in front of it (see top right). As the quasar varies in brightness, all four images vary too, but at slightly different times because the light of each image takes a different route around the lensing object. Astronomers can use the time difference between the variations of each image to estimate the distances to the quasar and the lensing galaxy. This provides a powerful way of determining the scale of the Universe independent of Cepheid variables and other methods which are open to calibration errors.

Careful analysis of the lensed images can reveal the distribution of dark matter in individual lensing galaxies and entire clusters. Gravitational lenses also make the images larger and brighter, like a natural magnifying glass, affording astronomers a better view of the stellar content of distant galaxies than would be possible with direct vision alone. The Hubble Space Telescope's ability to see many more examples of these cosmic mirages than is possible from the ground has transformed them from scientific curiosities to potentially powerful tools for probing the remote Universe.

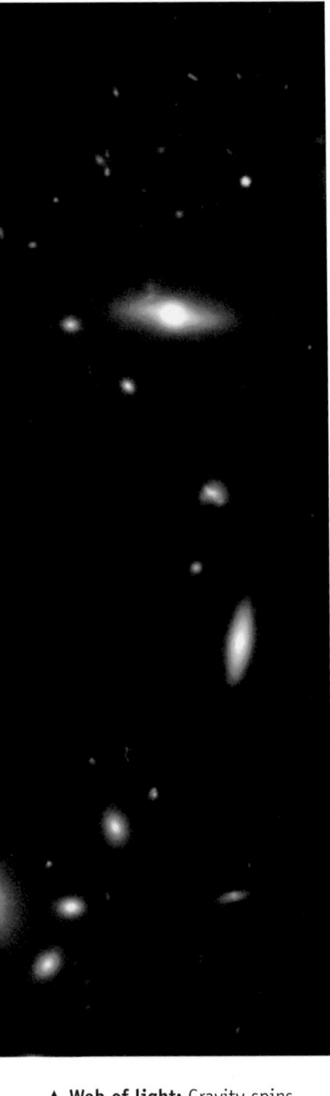

▲ **Web of light:** Gravity spins beams of light into a delicate web around members of a massive cluster of galaxies called Abell 2218, some 2 billion light years away in the northern constellation Draco. The slender crescents in this Hubble Space Telescope view are the distorted images of galaxies several billion light years beyond the lensing cluster. The colours yield clues to the ages and temperatures of stars within the lensed galaxies. Blue-white arcs show the existence of hot young stars, whereas the flaxen strands represent the combined light of stars of many ages.

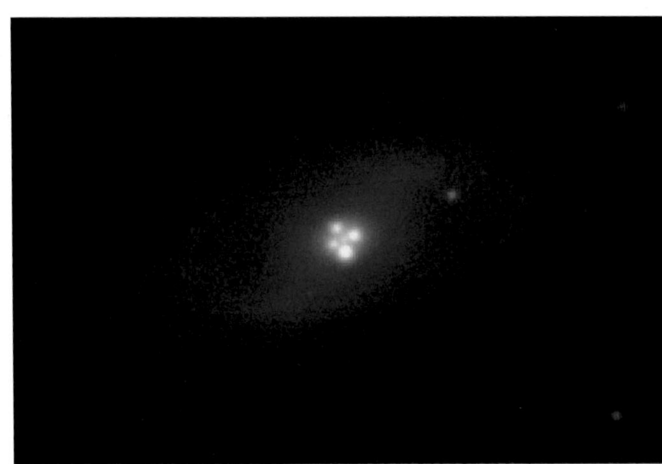

▶ **Einstein Cross:** Light from a quasar 8 billion light years away is split into four separate images by a gravitational lens. In this case, the lensing object is an ultra-massive black hole at the core of a spiral galaxy 20 times closer to us. The galaxy's core itself which contains the black hole can just be seen at the centre of the four cloned quasar images in this photograph from the WIYN (Wisconsin, Indiana. Yale, NOAO) telescope on Kitt Peak, Arizona.

▶ **In the loop:** The blue, loop-shaped objects in this Hubble Space Telescope photograph are multiple images of the same distant galaxy, gravitationally lensed by a foreground cluster of elliptical and spiral galaxies. One image is near the centre of the cluster; others are distributed in arcs around the cluster. Bright patches in the blue galaxy are young stars, while the dark core inside the ring is dust. The lensing cluster is 5 billion light years distant in the constellation Pisces, and the background galaxy is about twice as far away.

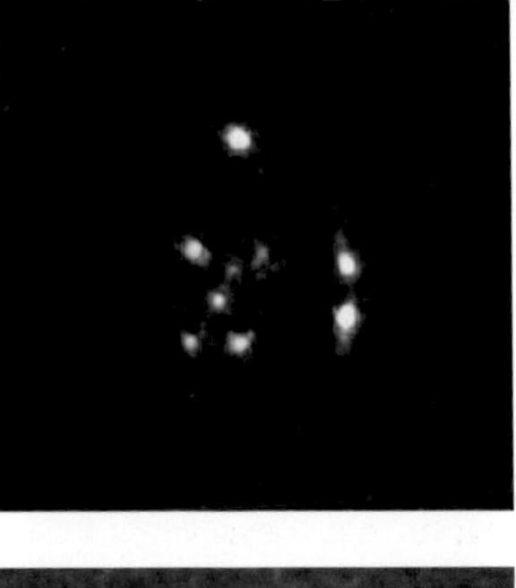

◀ **Six pack:** In one of the most complex interplays of light and gravity yet seen, the Hubble Space Telescope has detected a six-image gravitational lens in the northern constellation Boötes. The white objects are the images of the background galaxy, over 11 billion light years away; they are produced by a triangle of galaxies 4 billion light years closer, which appear orange. The lensed galaxy shows signs of a massive black hole at its core and has regions in which new stars are forming.

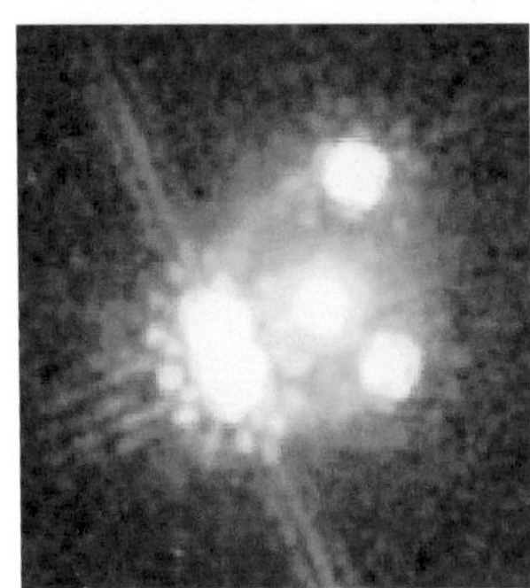

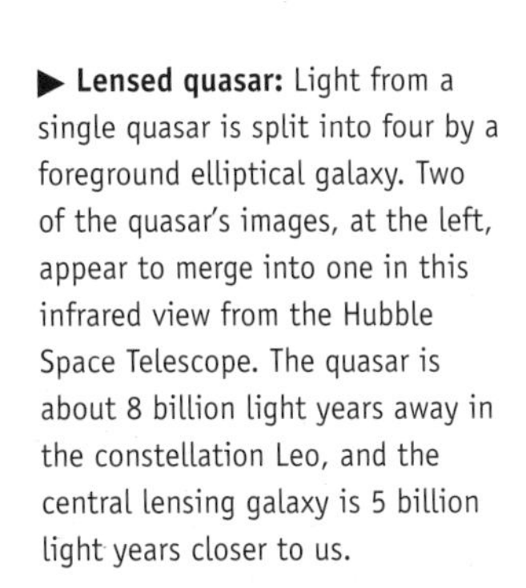

▶ **Lensed quasar:** Light from a single quasar is split into four by a foreground elliptical galaxy. Two of the quasar's images, at the left, appear to merge into one in this infrared view from the Hubble Space Telescope. The quasar is about 8 billion light years away in the constellation Leo, and the central lensing galaxy is 5 billion light years closer to us.

Looking deep into the Universe

Remote galaxies glow with ancient starlight in this Hubble Space Telescope view, known as the Hubble Deep Field, which reaches out through some 12 billion light years to the edge of the visible Universe. Covering a patch of sky only one-twelfth the diameter of the Moon, it was assembled from 342 individual exposures taken in December 1995, building up a picture of fainter and more distant objects than ever seen before. The area depicted is in the northern constellation Ursa Major, chosen because it contains no bright foreground stars or nearby galaxies to impede Hubble's view into the abyssal depths of time and space.

Scattered over the field of view is an assortment of some 3000 galaxies at various distances and stages of evolution. Extending the count from this small area to the entire sky, around 100 billion galaxies must be within Hubble's range. Among the distant galaxies in this image may be one that resembles our own home Galaxy as it appeared billions of years ago, before the Sun was born. In addition to recognizable spirals and ellipticals the Hubble Deep Field shows many others with unusual shapes, some of which may be so far off that we see them as they were during their formation, a billion years or so after the Big Bang. However, it is not immediately apparent from the image which galaxies are nearby and which are remote, so exhaustive follow-up studies of each one are being undertaken, both with telescopes on the ground and with Hubble.

Because the Universe is thought to look the same in all directions, the Hubble Deep Field should be representative of the far-off Universe as a whole, even though it covers only a tiny fraction of the sky. However, to check this assumption, Hubble took a second deep-field view nearly three years after the first in a completely different direction, this time in the southern constellation Tucana. The distribution of remote galaxies in these deep fields is used to test theories of the structure and evolution of the Universe.

To see farther off in space than these deep optical fields we must look in the infrared. This is because the expansion of the Universe stretches the light emitted by distant objects towards infrared wavelengths, an effect known as redshift. In 1998, the Hubble Space Telescope turned its newly installed infrared camera on part of the sky contained in the deep field image. The resulting image, shown in the inset, is thought to contain galaxies even farther away than those in the main optical image. However, their true nature can be confirmed only by larger telescopes of the future, notably Hubble's successor, the James Webb Space Telescope, which is specifically designed to observe in the infrared.

▶ **Postcards from the edge:** The Hubble Deep Field is a multi-exposure view of a small patch of sky near the familiar star pattern of the Plough (or Big Dipper). To obtain this detailed view, the Hubble Space Telescope photographed the same small area of sky repeatedly over 10 consecutive days, each exposure lasting between 15 and 40 minutes. The resulting 342 frames were stacked together to reveal objects too faint, and details too delicate, to see in a single exposure. Most of the objects seen here are galaxies and their building blocks, extending all the way to the visible horizon of the Universe. A handful of stars in our own Galaxy also appear in the image – they are the rounded images with spikes, such as at upper right. During the entire series of exposures, the Hubble Space Telescope orbited the Earth 150 times but remained precisely pointed at its target throughout. The white box shows the area of the infrared view below.

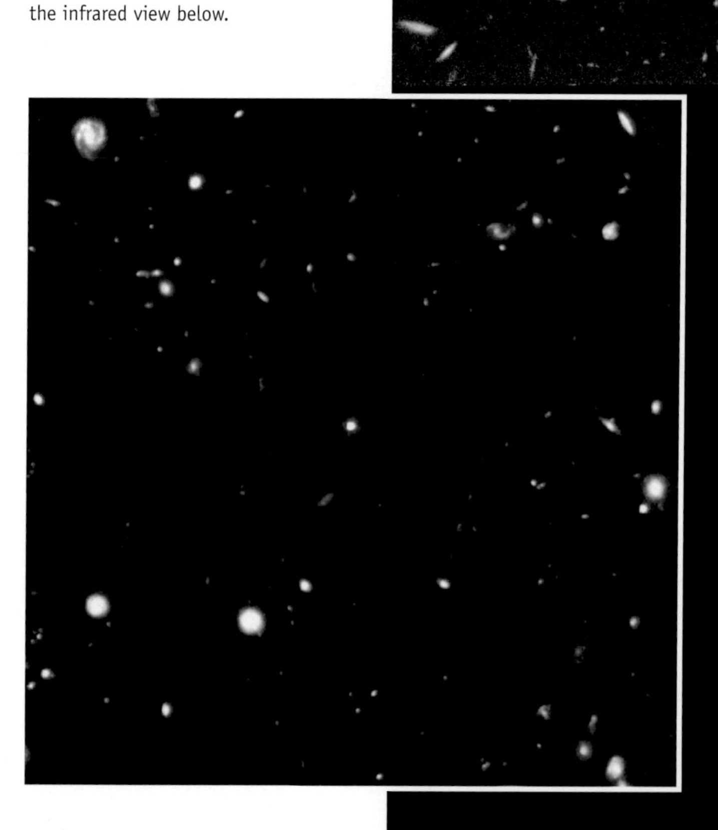

▶ **Into the red:** An infrared view of part of the Hubble Deep Field, indicated by the white box on the main image. Over 300 faint galaxies are captured on this 36-hour exposure, some not seen in visible light. The infrared view appears less sharp because infrared wavelengths are longer than visible ones.

The birth and fate of the Universe

For the first 380,000 years of its existence, the Universe was filled with an inferno of matter and energy disgorged by the Big Bang. As the Universe expanded the temperature of this incandescent broth steadily dropped. Once it had cooled below 3000°C, protons could link up with electrons to form the first atoms – mostly hydrogen gas but with about 10% helium and a smattering of other light elements. At that stage the Universe, previously clouded by free-roaming electrons, became transparent; light and other forms of radiation could at last travel through it without interference. That primeval light still permeates the Universe, albeit redshifted by the cosmic expansion into the short-wavelength radio region of the spectrum where it is detected as the so-called cosmic microwave background radiation. Observing this radiation is the closest we can get to a direct view of the Big Bang itself.

The existence of the cosmic background radiation is direct evidence that the Universe was born in the fiery furnace of a hot Big Bang. Since then, the expansion of the Universe has cooled the radiation to a temperature of a mere 2.7 degrees above absolute zero, a thousand times less than when it was emitted. Although the radiation is almost uniform across the entire sky, there are gentle contours in its temperature from place to place. These highs and lows, amounting to no more than about 0.0001°C, result from variations in density of the gas from the Big Bang fireball, the denser regions appearing cooler than average and the less dense regions being warmer. They are the first sketchy outlines of the clusters of galaxies that were to grow by gravitational collapse over the next billion years of the Universe's history.

NASA's Cosmic Background Explorer (COBE) satellite first detected these subtle imprints in the cosmic microwave background in 1992. A more detailed view was provided in 2003 by the Wilkinson Microwave Anisotropy Probe, WMAP (see top illustration on facing page). Limited areas of sky have also been observed in even greater detail by Earth-bound and balloon-borne experiments. By combining the results from all these sources, astronomers have been able to draw far-reaching conclusions about the Universe and its contents, including dating the Big Bang to 13.7 billion years ago, in line with earlier estimates from the Hubble Space Telescope.

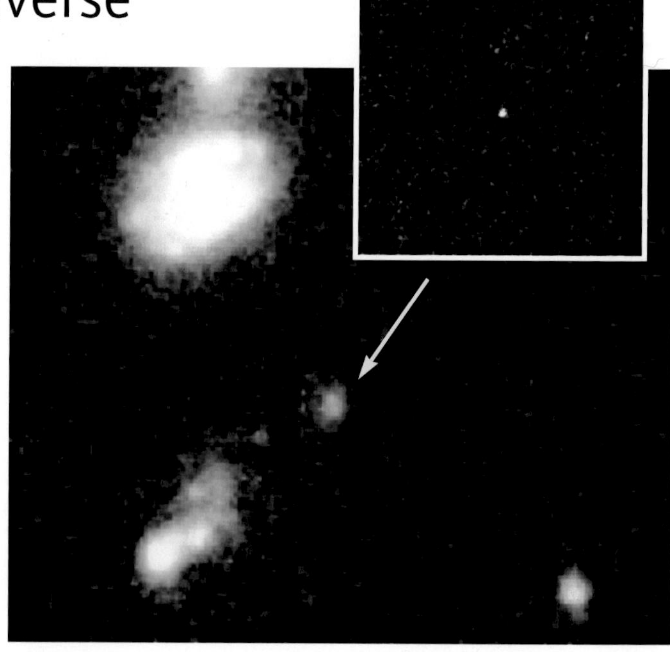

▲ **Ancient supernova:** This distant galaxy (arrowed) on the Hubble Deep Field survey was observed again after an interval of two years, when a supernova was found to have exploded in it. In the smaller picture, the light of the galaxy has been removed to leave only the supernova. It is the most distant supernova ever seen, 10 billion light years away. By comparing the brightness of such remote supernovae with those closer to us, astronomers have deduced that the rate of expansion of the Universe is speeding up.

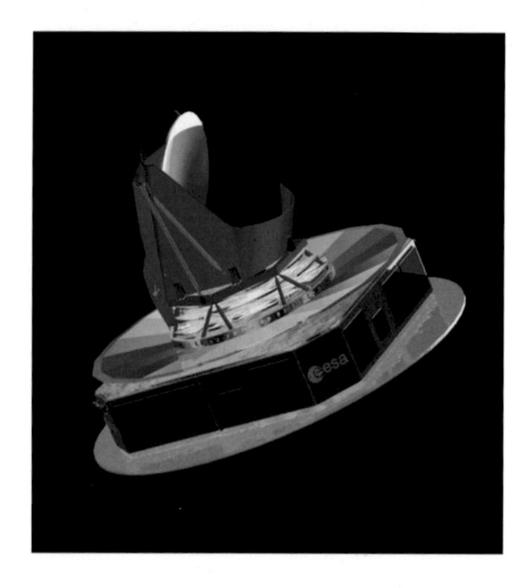

◀ **Planck:** Named after the German physicist Max Planck, this European Space Agency spacecraft will measure fluctuations in the cosmic background radiation in more detail and with greater sensitivity than ever before. Planck is due for launch in 2007.

Astonishingly, it turns out that only about 4% of the mass in the Universe resides in the familiar type of matter that makes up stars, nebulae, planets and us. At least five times as much mass is in a still-unknown form, termed dark matter. Among the possible candidates for dark matter are heavy subatomic particles that have so far eluded discovery. Although invisible and unknown, dark matter is not irrelevant, for without its gravitational effects the first galaxies would not have been able to form.

Most astoundingly of all, nearly three-quarters of the total mass of the Universe is now thought to be in the form of so-called "dark energy", a mysterious force associated with space itself. Observations of distant supernovae indicate that, rather than slowing down due to gravity's drag, the expansion of the Universe is actually speeding up. The repulsive effect of dark energy started to override the attraction of gravity when the Universe was about half its current age, boosting the rate of expansion. Hence it seems that the fate of the Universe is to expand forever, impelled by the force of dark energy. Understanding the nature of dark energy will be one of the prime goals of cosmology in the years to come.

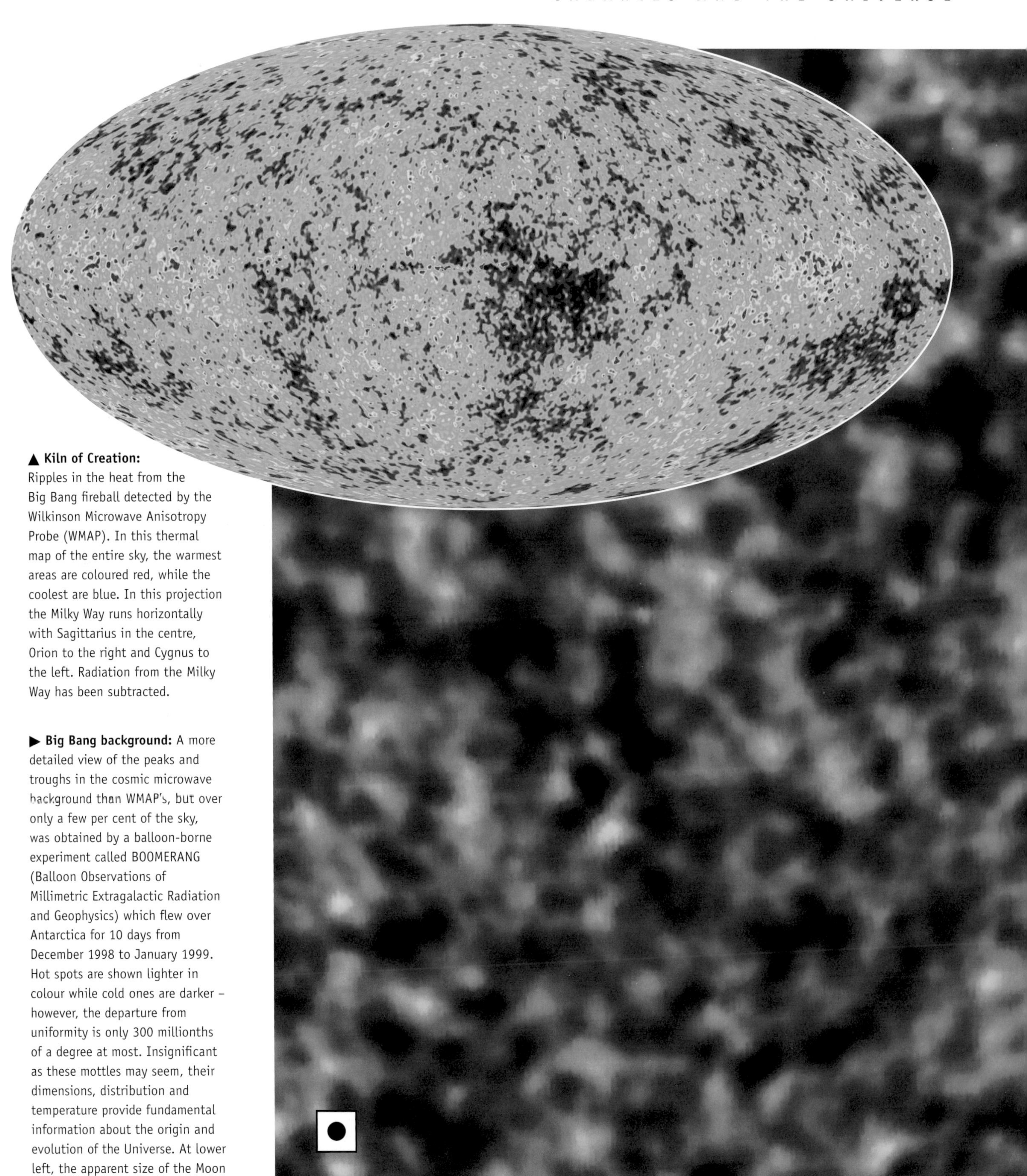

▲ **Kiln of Creation:** Ripples in the heat from the Big Bang fireball detected by the Wilkinson Microwave Anisotropy Probe (WMAP). In this thermal map of the entire sky, the warmest areas are coloured red, while the coolest are blue. In this projection the Milky Way runs horizontally with Sagittarius in the centre, Orion to the right and Cygnus to the left. Radiation from the Milky Way has been subtracted.

▶ **Big Bang background:** A more detailed view of the peaks and troughs in the cosmic microwave background than WMAP's, but over only a few per cent of the sky, was obtained by a balloon-borne experiment called BOOMERANG (Balloon Observations of Millimetric Extragalactic Radiation and Geophysics) which flew over Antarctica for 10 days from December 1998 to January 1999. Hot spots are shown lighter in colour while cold ones are darker – however, the departure from uniformity is only 300 millionths of a degree at most. Insignificant as these mottles may seem, their dimensions, distribution and temperature provide fundamental information about the origin and evolution of the Universe. At lower left, the apparent size of the Moon is shown to the same scale.

Index

Page numbers in **bold** refer to illustrations.

ABOUT THE AUTHOR:

Ian Ridpath is an author and broadcaster on astronomy and space research. He is the author of three standard observing guides to the night sky for amateur astronomers: the *Collins Pocket Guide to Stars & Planets* (in the US, the *Princeton Field Guide to Stars & Planets*), Collins Gem *Stars*, and *The Monthly Sky Guide*. He is editor of *Norton's Star Atlas* and of the authoritative *Oxford Dictionary of Astronomy*. His star show Planet Earth ran at the London Planetarium for two years.
www.ianridpath.tk

Picture credits

1 (left) Steve Lee, Jim Bell, Mike Wolff/NASA
1 (centre) NASA/ESA
1 (right) Nigel Sharp/AURA/NOAO/NSF
2–3 T.A. Rector (NOAO/AURA/NSF) and Hubble Heritage Team (STScI/AURA/NASA)
4 Lick Observatory
5 USGS
6–7 Nik Szymanek
8 (top) European Southern Observatory
8 (bottom) John Hill
9 (top) Tom Sebring/AURA/Gemini Observatory/ NOAO/NSF
9 (left) Gemini Observatory/AURA/NOAO/NSF
9 (bottom) Gemini Observatory/NSF/ University of Hawaii Institute for Astronomy
10 NASA
10 (inset l) NASA
10 (inset r) NASA
11 NASA
12–13 Images NASA; montage Ian Ridpath/ HarperCollins
14 Reto Stockli, Alan Nelson, Fritz Hasler/ NASA/GSFC
15 NOAA
16–17 Craig Mayhew, Robert Simmon/NASA/ GSFC/DMSP
16 (bottom) NASA
17 (bottom) NASA/GSFC
18 (centre) NASA
18 (bottom) NASA
19 (top) NASA
19 (bottom l) NASA
19 (bottom r) NASA
20 (bottom l and r) USGS
21 Lick Observatory
22 (top) NASA/Scan by Kipp Teague
22 (bottom l) NASA
22 (bottom r) NASA/Scan by Kipp Teague
23 (top) NASA/Scan by Kipp Teague
23 (bottom) NASA/Scan by Kipp Teague
24 (top) NASA/Scan by Kipp Teague
24 (bottom) NASA/Scan by Kipp Teague
25 (top) NASA/Scan by Kipp Teague
25 (bottom) NASA/Scan by Kipp Teague
26 (top) NASA/Scan by Kipp Teague
26 (centre) NASA/Scan by Kipp Teague
26 (bottom) NASA/Scan by Kipp Teague
27 (top) Juan Carlos Casado
27 (bottom) Fred Espenak
28 (bottom) NASA/JPL/Northwestern University
29 (top l) NASA/JPL/Northwestern University
29 (top r) NASA/JPL/Northwestern University
29 (centre) NASA/JPL/Northwestern University
29 (bottom) ICSTARS Astronomy
30 (bottom l) L. Esposito/NASA
30 (bottom r) NASA/Ames
31 NASA/JPL/MIT
32 (top) NASA/JPL
32 (bottom) USSR Academy of Sciences/Brown University
33 (top l) NASA/JPL
33 (centre l) NASA/JPL
33 (centre r) NASA/JPL
33 (bottom) NASA/JPL/Ian Ridpath
34 (bottom) NASA/JPL
35 Steve Lee, Jim Bell, Mike Wolff/NASA
36 (top) Image: Michael Caplinger and Michael Malin, Malin Space Science Systems. Data: NASA/JPL/MSSS/MOLA Science Team
36 (bottom) NASA/USGS
37 (top) NASA/USGS
37 (centre) NASA/USGS
37 (bottom) NASA/MOLA Science Team
38–39 NASA/JPL
38 (centre) NASA/JPL
38 (bottom) NASA/JPL
39 (all) NASA/JPL
40 (top) NASA/MOLA Science Team
40 (bottom) NASA/JPL/Malin Space Science Systems
41 (top) NASA/MOLA Science Team
41 (bottom) NASA/JPL/Malin Space Science Systems
42 (top) NASA/GSFC/JHUAPL/NLR
42 (bottom l) NASA/JHUAPL
42 (bottom r) NASA/JHUAPL
43 (top) NASA/JPL
43 (centre r) NASA/JPL
43 (bottom l) V.L. Sharpton/LPI
44 (bottom) NASA/JPL
45 NASA/JPL/University of Arizona
46 (top) NASA/JPL/Lick Observatory
46 (bottom l) NASA/JPL/University of Arizona
46 (bottom r) NASA/JPL/LPL
47 (top l and r) NASA/JPL/University of Arizona
47 (centre) NASA/JPL/University of Arizona
47 (bottom l) NASA/JPL/University of Arizona
47 (bottom r) NASA/JPL/DLR
48 (top l) NASA/JPL/DLR
48 (centre r) NASA/JPL/DLR
48 (bottom l) NASA/JPL/Brown University
48 (bottom r) NASA/JPL/Brown University
49 (top l) NASA/JPL/DLR
49 (centre r) NASA/JPL
49 (centre l) NASA/JPL/Arizona State University
49 (bottom) NASA/JPL
50 (bottom) Reta Beebe (NMSU)/D. Gilmore, L. Bergeron (STScI)/NASA
51 (main) NASA/JPL/USGS
51 (inset) NASA/JPL
52 (centre) NASA/JPL
52 (bottom l) Erich Karkoschka (University of Arizona)/NASA
52 (bottom r) NASA/JPL
53 (top r) Peter H. Smith (University of Arizona)/NASA
53 (centre l) NASA/JPL
53 (centre) NASA/JPL
53 (centre r) NASA/JPL
53 (bottom l) NASA/USGS
54 (bottom l) NASA/JPL
54 (centre r) NASA/USGS
54 (bottom r) NASA/JPL
55 (main and inset) Erich Karkoschka (University of Arizona)/NASA
56 (bottom l) NASA/JPL
56 (centre r) NASA/USGS
57 (top) NASA/JPL
57 (bottom r) Lawrence Sromovsky (University of Wisconsin-Madison)/NASA
58 (bottom) Alain Doressoundiram (Observatoire de Paris)/Christian Veillet (CFH Institute)
59 (top) R. Albrecht/NASA
59 (inset) Eliot Young (SwRI)/NASA
60 Lorenzo Comolli
61 (top) Paul Sutherland
61 (bottom) H.U. Keller/Max-Planck-Institut für Aeronomie
62 (top) SOHO-LASCO consortium/ ESA/NASA
62 (centre r) University of Hawaii
62 (bottom r) NASA/Harold Weaver (JHU)/ HST Comet LINEAR Investigation Team
63 (top) H.A. Weaver, T.E. Smith/NASA
63 (centre r) H. Hammel (MIT)/NASA
63 (bottom) NASA/JPL/Brown University
64–5 Juan Carlos Casado
66 (bottom) Mees Solar Observatory, University of Hawaii
67 SOHO-EIT consortium/NASA/ESA
68 (top r) Louis Strous (Lockheed-Martin)/ Stanford/NASA/ESA
68 (centre r) Sam Freeland (Lockheed-Martin)/ ISAS/NASA
68 (bottom) Kiepenheuer-Institut für Sonnenphysik
69 (top) SOHO-LASCO consortium/ NASA/ESA
69 (centre r) NASA/JSC
70 Fred Espenak
71 Fred Espenak
72–3 ICSTARS Astronomy
73 (top) ICSTARS Astronomy
74 Copyright © Anglo-Australian Observatory. Photograph by David Malin
75 (centre) European Southern Observatory
75 (bottom l) C.R. O'Dell, S.K. Wong (Rice University)/NASA
75 (bottom r) K.L. Luhman (Harvard-Smithsonian Center for Astrophysics)/G. Schneider, E. Young, G. Rieke, A. Cotera, H. Chen, M. Rieke, R. Thompson (Steward Observatory)/NASA
76 (top) The Hubble Heritage Team (STScI/AURA)/ESA/NOAO/NASA
76 (bottom) Hui Yang (University of Illinois)/ NASA
77 (main) Jeff Hester, Paul Scowen (Arizona State University)/NASA
77 (inset) Copyright © Anglo-Australian Observatory. Photograph by David Malin
78 (top) Copyright © Anglo-Australian Observatory. Photograph by David Malin
78 (inset) Jeff Hester (Arizona State University)/NASA
78 (bottom r) The Hubble Heritage Team (STScI/AURA)/NASA
79 (main) T.A. Rector (NRAO/AUI/NSF and NOAO/AURA/NSF) and B.A. Wolpa (NOAO/AURA/NSF)
79 (inset) NASA and the ACS Science Team
80 (left) Chris Burrows (STScI)/NASA
80 (centre top) M. McCaughrean (MPIA)/ C.R. O'Dell (Rice Univ.)/NASA
80 (centre r) J. Bally (University of Colorado)/ H. Throop (SwRI)/C.R. O'Dell (Vanderbilt University)/NASA
80 (centre l) M. McCaughrean (MPIA)/C.R. O'Dell (Rice Univ.)/NASA
80 (bottom) C. Burrows, J. Krist (STScI)/NASA
81 (top) Copyright © Anglo-Australian Observatory. Photograph by David Malin
81 (bottom) The Hubble Heritage Team (AURA/STScI)/NASA
82 (centre) Nigel Sharp/AURA/NOAO/NSF
82 (bottom) Nigel Sharp, Mark Hanna/ AURA/NOAO/NSF
83 (main) Copyright © Anglo-Australian Observatory/Royal Observatory, Edinburgh. Photograph from UK Schmidt plates by David Malin
83 (inset) The Hubble Heritage Team (STScI/AURA)/NASA
84 (top) Martino Romaniello (European Southern Observatory)/ESA/NASA
84 (bottom) Copyright © Anglo-Australian Observatory. Photograph by David Malin
85 Wolfgang Brandner (JPL/IPAC)/Eva K. Grebel (Univ. Washington)/You-Hua Chu (Univ. Illinois Urbana-Champaign)/NASA
86 (centre l) Raghvendra Sahai, John Trauger (JPL)/WFPC2 science team/NASA
86 (bottom r) J.P. Harrington, K.J. Borkowski (University of Maryland)/NASA
87 (main and inset) NASA, NOAO, ESA, the Hubble Helix Nebula Team, M. Meixner (STScI), and T.A. Rector (NRAO)
88 (top) Hubble Heritage Team (AURA/STScI/NASA)
88 (centre) The Hubble Heritage Team (STScI/AURA)/ESA/NASA
88 (bottom) The Hubble Heritage Team (STScI/AURA)/NASA
89 (top) A. Fruchter and the ERO Team (STScI)/NASA
89 (bottom) Garrelt Mellema (Leiden University)/ESA/NASA
90 (left) Don F. Figer (UCLA)/NASA
90 (right) Jon Morse (University of Colorado)/ NASA
91 (top l and r) Copyright © Anglo-Australian Observatory. Photograph by David Malin
91 (centre) P. Challis (CfA)/NASA
92 (top r) J. Hughes (Rutgers)/SAO/CXC/NASA
92 (left) Copyright © Malin/IAC/RGO. Photograph by David Malin
93 (main) Copyright © Malin/Pasachoff/ Caltech. Photograph by David Malin
93 (inset) The Hubble Heritage Team (STScI/AURA)/NASA
94–95 Copyright © Anglo-Australian Observatory. Photograph by David Malin
96–97 Digital Sky LLC
97 (bottom) COBE Science Working Group/ NASA/GSFC
98 Copyright © Anglo-Australian Observatory. Photograph by David Malin
99 (main) Copyright © Anglo-Australian Observatory/Royal Observatory, Edinburgh. Photograph from UK Schmidt plates by David Malin
99 (inset) Copyright © Anglo-Australian Observatory/Royal Observatory, Edinburgh. Photograph from UK Schmidt plates by David Malin
100 (all) John Dubinski (University of Toronto)
101 © Caltech/David Malin
102 Copyright © Malin/IAC/RGO. Photograph by David Malin
103 (main) Copyright © Anglo-Australian Observatory. Photograph by David Malin
103 (inset) Copyright © Anglo-Australian Observatory. Photograph by David Malin
104 (top) The Hubble Heritage Team (STScI/AURA)/NASA
104 (bottom) Peter Bunclark and Nik Szymanek/Isaac Newton Group of Telescopes
105 (top) The Hubble Heritage Team (STScI/AURA)/NASA
105 (bottom) The Hubble Heritage Team (STScI/AURA)/NASA
106 (top) Copyright © Anglo-Australian Observatory. Photograph by David Malin
106 (bottom) Copyright © Anglo-Australian Observatory. Image by S. Lee, C. Tinney, D. Malin
107 (top) European Southern Observatory
107 (inset c) John Trauger (Jet Propulsion Laboratory)/NASA
107 (inset r) C. Marcella Carollo (Johns Hopkins University and Columbia University)/ ESA/NASA
107 (bottom l) D. Maoz (Tel-Aviv University and Columbia University)/ESA/NASA
108 (top) Roeland P. van der Marel (STScI), Frank C. van den Bosch (Univ. of Washington)/ NASA
108 (bottom) The Hubble Heritage Team (STScI/AURA)/NASA
109 (top) The Hubble Heritage Team (STScI/AURA)/NASA
109 (centre l) Nigel Sharp/NOAO/AURA/NSF
109 (lower r) Jayanne English (University of Manitoba), Sally Hunsberger (Pennsylvania State University), Zolt Levay (Space Telescope Science Institute), Sarah Gallagher (Pennsylvania State University), Jane Charlton (Pennsylvania State University)/NASA
110 Jeffrey Newman (Univ. of California at Berkeley)/NASA
111 (top) High-Z Supernova Search Team/NASA
111 (bottom) Wendy L. Freedman (Observatories of the Carnegie Institution of Washington)/NASA
112 (top, both) Kirk Borne (Raytheon and NASA GSFC)/Luis Colina (Instituto de Fisica de Cantabria, Spain)/Howard Bushouse and Ray Lucas (STScI)/NASA
112 (bottom) The Hubble Heritage Team (STScI)/NASA
113 (main) Copyright © Anglo-Australian Observatory. Photograph by David Malin
113 (inset) E.J. Schreier (STScI)/NASA
114 (top) Copyright © Anglo-Australian Observatory. Image by S. Lee, C. Tinney, D. Malin
114 (bottom) Brad Whitmore (STScI)/NASA
115 (top) Kirk Borne (STScI)/NASA
115 (bottom) The Hubble Heritage Team (AURA/STScI)/NASA
116 (top) Copyright © Subaru Telescope, National Astronomical Observatory of Japan. All rights reserved
116 (bottom) NASA and the ACS Science Team
117 Javier Méndez and Nik Szymanek/ Isaac Newton Group of Telescopes
118 (left) STScI/NASA
118 (right, all) John Bahcall (Institute for Advanced Study, Princeton)/Mike Disney (University of Wales)/NASA
119 Andrew S. Wilson (University of Maryland)/Patrick L. Shopbell (Caltech)/ Chris Simpson (Subaru Telescope)/Thaisa Storchi-Bergmann and F.K.B. Barbosa (UFRGS, Brazil)/Martin J. Ward (University of Leicester)/NASA
120–121 A. Fruchter and the ERO Team (STScI, ST-ECF)/NASA
121 (top r) J. Rhoads, S. Malhotra, I. Dell'Antonio (NOAO)/WIYN/NOAO/NSF
121 (centre r) W.N. Colley and E. Turner (Princeton University)/J.A. Tyson (Bell Labs, Lucent Technologies)/NASA
121 (below centre) David Rusin (University of Pennsylvania)/STScI/ESA/NASA
121 (bottom) Christopher D. Impey (University of Arizona)/NASA
122–123 Robert Williams and the Hubble Deep Field Team (STScI)/NASA
122 (inset) Rodger I. Thompson (University of Arizona)/NASA
124 (top and inset) Adam Riess (STScI)/NASA
124 (bottom) ESA
125 (top) NASA/WMAP Science Team
125 (main) Boomerang collaboration